SURVEYING MEASUREMENTS
and their Analysis

by R. B. Buckner

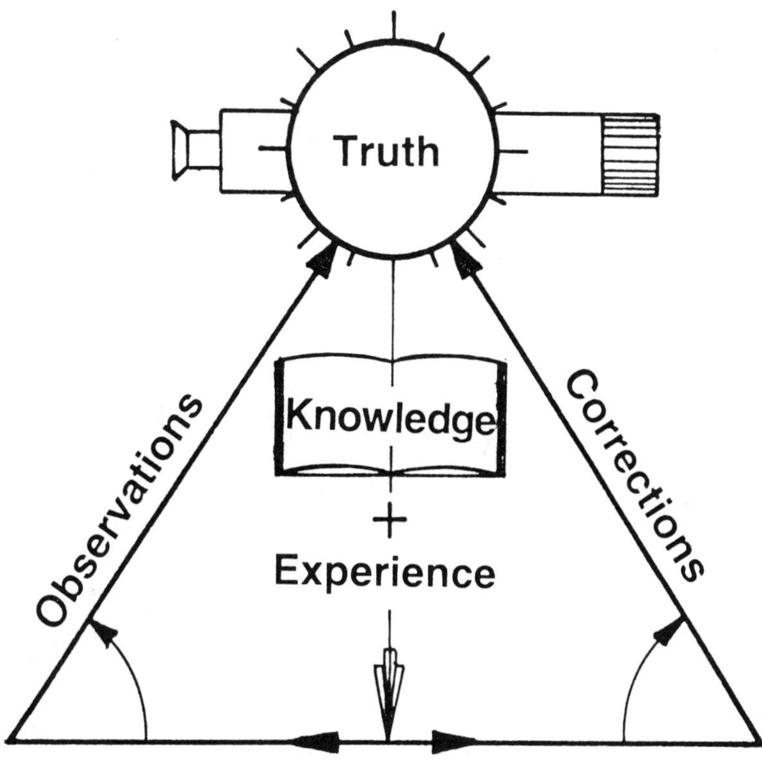

The 'Truth' is equal to one's initial observations plus the corrections discovered through added experience or knowledge.

SURVEYING MEASUREMENTS AND THEIR ANALYSIS

by

R. B. BUCKNER, Ph.D.

Landmark Enterprises
10324 Newton Way
Rancho Cordova, California 95670

First Edition

Second Printing, March 1986
Third Printing, May 1991

International Standard Book Number
ISBN 0-910845-11-5

MANUFACTURED IN THE UNITED STATES OF AMERICA

Typography by TOTAL GRAPHIC SERVICES
Loomis California

Preface

The main purpose of this book is to explain the nature of surveying measurement to surveying technicians, professional surveyors and students. It has the goal of helping to develop these individuals into true experts of measurement. Those who study this book should emerge with an appreciation for the limitations of people and instruments in the world of measurement. They should be able to control errors in measurements, design measurement systems, establish standards of practice, and enjoy the feeling of being knowledgeable in the key aspects of the surveying profession. The art and the science of measurement merge in this book.

Other objectives of this book are to explain the complexities of measurement to civil engineers and others who use and depend upon surveying measurements for part of their work, to mathematics teachers at all levels of teaching, and to others interested in the mysteries of measurement. By learning about the nature of measurement, others should be able to better appreciate the role and expertise of the surveyor. After carefully considering the difference between measurements and counts, mathematics teachers will better understand that dealing with numbers when teaching geometry and trigonometry requires consideration for significant figures and round-off errors. Hopefully, they will be able to convey these concepts to students. This should surely lead to a more enlightened general public in future generations. Chapters 1-3, sections 4.1-4.3, and most of Chapter 9 are especially suited for readers seeking an appreciation for, but not complete expertise in measurement science.

This book differs from other surveying texts. Available books on basic surveying explain techniques and methods adequately, but contain scant discussion of the true nature of measurement. Such books contain only brief definitions of measurement terms with perhaps a few examples on errors, but the art and science of measurement remains undeveloped. On the other extreme are texts on adjustment computations and advanced theory which also merely define measurement terms and give brief examples. This book fills the gap between the elementary books on methods and the ones on adjustments. The scope includes analysis of results, estimation of uncertainty, and design of specifications, which is often called "pre-analysis" in the few texts where it is briefly considered.

This book is intended to serve as a text in colleges offering degree programs in surveying, either at the baccalaureate or associate degree level. It is also intended to serve as a reference for surveying practitioners, to enable them to return to the basics, possibly revealing truths, the misunderstanding of which may have caused them frustration in analyzing measurements and developing measurement specifications and standards. Although it is intended to serve as a single text in beginning surveying courses, students should have some differential calculus in order to completely follow derivations after Section 4-4. The material up to Section 4-5 can be understood by readers having very little math background. Field exercises should be included, either in a prior course or during presentation of Chapter 3, so that the process of measurement and occurrence of errors is appreciated. Traverse computations and other plane surveying mathematics should be included, either in a prior course or in conjunction with study of Chapter 8, so that the material in that chapter can be fully developed and expanded.

For surveying practitioners and technicians with field and computation background this book should be sufficient as a sole self-study text on analyzing measurements. Even without the calculus background, the study of this text should help develop analysis skills and a better appreciation of the art and science of measurement since alternate, simplified methods have been introduced in Section 4-6 and can be used in lieu of calculus in many situations.

Part A develops the nature of measurement. Readers with a good understanding of instruments and elementary surveying may wish to merely skim Chapters 1 and 3, but should thoroughly study Chapters 2 and 4. Part B demonstrates the theories explained in Chapter 4 by applying the theory to analysis of common measurement situations. Part C is meant to extend the depth of coverage into professional practice, hopefully instilling an appreciation of the need for expertise in measurement. The fine points of Chapter 9 should stimulate interest in the profession and an appreciation of its unique and specialized nature. The main theme of the book ends with Chapter 9. Chapter 10, although part of the section on controlling measurements, is an introduction to another important aspect of measurement and serves to illustrate that the subject of measurement analysis does not end with the theme developed herein. The last chapter is intended to stimulate interest in further study.

References at the end of chapters are intended to serve as supplementary reading. They do not duplicate coverage herein but generally represent related material which may help to either clarify certain points where individual background is lacking, or in some instances, provide connections to more advanced material.

The metric system is used in about half of the text examples and problems. The English system receives equal use, however, since land surveyors, as well as engineering surveyors will need to continue to work with these units in the foreseeable future. Surveyors, as experts in measurement, should be thoroughly familiar and comfortable with all such distance, angle, temperature, area, and weight units. The treatment herein is intended to *develop* familiarity in metrics in a way that is comfortable and acceptable.

The author wishes to acknowledge the help of George Schweitzer, a former student, who checked the examples and offered other suggestions, and to Dr. David Knowles, Paul De Nu and Porter McDonnell who reviewed the portions of the manuscript for technical suitability. Ideas were also gained through association with others interested in the subject, particularly Dr. Urho Uotila and Professor David Gibson. Their insights surface here and there in the book. I am especially grateful to Roy Minnick, who gave me encouragement during the publication of the book. The students who appear as rodmen in Chapter 3 are Bassey Ukim and Mark Ellis. Both men are graduates of the Bachelor of Science in Surveying degree program at the Ohio State University.

This book is dedicated to Dr. Gordon Gracie, my first surveying instructor, who instilled in me an appreciation for measurements and their uncertainty.

Dec. 1981

CONTENTS

4. SURVEY MEASUREMENT THEORY

B — ANALYZING MEASUREMENTS 119

5. ANALYSIS OF DISTANCE MEASUREMENTS

6. ANALYSIS OF ANGLE MEASUREMENTS

7. ANALYSIS OF ELEVATION MEASUREMENTS

8. ANALYSIS OF INDIRECT MEASUREMENTS

C — CONTROLLING MEASUREMENTS 235

9. THE SURVEYOR AS EXPERT MEASURER

10. INTRODUCTION TO ADJUSTMENTS OF MEASUREMENT DATA

List of Illustrations

Figure		Page

List of Tables

Section A
UNDERSTANDING MEASUREMENTS

1. INTRODUCTION TO SURVEYING MEASUREMENTS
2. THE WORLD OF MEASUREMENT
3. SURVEYING INSTRUMENTS AND THEIR USE
4. SURVEY MEASUREMENT THEORY

CHAPTER 1

INTRODUCTION TO SURVEYING MEASUREMENTS

1-1 DEFINITION AND FUNCTION OF SURVEYING

One reference states the definition of surveying as "the science and art of making such measurements as are necessary to determine the relative position of points above, on, or beneath the surface of the earth, or to establish such points in a specified position" (Ref.6). Another textbook gives it as "the science and art of determining relative positions of points above, on, or beneath the surface of the earth, or establishing such points" (Ref. 3).

These definitions are similar and typify what is given in many surveying textbooks. Surveying, as can be seen from these definitions, is involved with measurements. These measurements are not confined to the land surface. A surveyor makes measurements over water and through space. Surveyors measure above and below the earth as well as on its surface.

The above definitions indicate that surveying is both a science and an art. It is often difficult for the beginner and sometimes for the experienced professional to comprehend which portion of surveying is science and which is art because the two merge into one function and into a technical and professional process which is accomplished without attempting to sort out the difference. The science portion is essentially that which involves mathematics, physics, and natural sciences. The art of surveying is the part that involves judgment and logical thinking. The art usually takes much longer for a surveyor to develop as an individual. Examples of the art of surveying would be a surveyor's arrangement of field notes, planning the steps of a field survey, proper selection of equipment, analysis of data, and making correct decisions.

It should be clear that surveying is a science. As will be seen, mathematics, particularly algebra and trigonometry, are applied in most surveying problems. The portion which is art is a little more subtle. But, the student should attempt to develop this part in order to eventually become a well rounded surveyor, capable of making accurate measurements and calculations. Development of the art of surveying often makes the difference between one becoming a professional surveyor or remaining at a technician level.

The last part of the definitions given above alludes to the fact that surveyors are faced with two fundamental problems: (1) to determine relative positions between existing points, and (2) to establish new points in a specified position. This is important to understand for reasons which will be more obvious when error analysis is discussed later in the book.

An example of the first of these problems is measuring the distance between two property monuments already set in the ground, where the surveyor takes a reading from a tape and thereby determines the existing distance. But, if only one marker exists and the surveyor wishes to establish a second one at a specified distance from the first, this is an example of the second fundamental problem, sometimes called "laying out."

The above definitions of surveying are rather simplified. The functions of modern-day surveyors are many and varied, as will be outlined later in this chapter and throughout the book. In fact, the *professional* surveyor, even though he or she would be competent to do so, may not actually make very many measurements. The measuring itself is often done by surveying *technicians* and the professional surveyor analyzes the data, evaluates the errors, makes decisions, directs draftsmen and computers as to how to plot and manipulate the data, selects equipment, and supervises several surveying projects simultaneously.

The American Society for Engineering Education has defined a *technician* as one who works in the occupational area between the craftsman and the professional. Technicians differ from craftsmen in their knowledge of theory and methods and differ from professionals in that their role is one of supporting the professional activities. Although there are many types of surveyors, there is a fairly clear distinction between those who are professional surveyors and those who are surveying technicians. The difference between the technician and the professional might be illustrated by a comparison with the medical field. When one has a thorough physical examination, several technicians collect the data (measurements). They are trained in their specialty and know how to use the equipment, take the measurements, and plot or record the readings. But the doctor analyzes the data, evaluates the readings, makes diagnoses, and prescribes remedies if any. The technicians in this or any field play an extremely important role. Their assistance is invaluable because it frees the professional to do other important things. Technicians may or may not become professionals in surveying. Becoming a professional depends on development of the art of surveying and depth of education in the science of surveying. Passing a professional licensing examination does not necessarily make one a professional.

Surveyors and surveying technicians work both outside and inside. Many prefer year-round outdoor work. Most like outdoor work to some extent. The work consists of the field measurements already discussed and to be detailed later. It also consists of measurements made indoors. Photogrammetric measurements are usually made indoors, for example. All plotting and drafting requires some scaling, which is usually performed indoors. The surveyor and surveying technician are never far away from measurements of some kind. In addition, surveyors and techncians are involved with calculations related to the measurements. The manipulation and analysis of the data requires basic skills in trigonometry and algebra in particular. More complex problems can only be solved using statistics, calculus, matrix algebra, or spherical trigonometry. Computer programming is also employed. Surveyors should also be skilled in plotting, arranging notes and calculations neatly, and other aspects of graphical communication.

It can be seen that the definition and function of surveying involves measurements, primarily, but also involves mathematics and other skills. The surveying technician makes measurements. The professional surveyor also may make measurements and analyze them as well. Among other roles, the surveyor (technician and professional) supports the civil engineer, architect, real estate broker and others with surveying services. Surveying technicians may work directly under either engineers or professional surveyors. Likewise, professional surveyors and engineers may work together as a team. A person may become educated and licensed as both an engineer and a surveyor. The engineer must understand the function of surveying and, unless he or she becomes thoroughy educated in this profession, must know when to employ expert surveying technicians or consult with professional surveyors for assistance and support.

This book is written for both surveying technicians and professional surveyors, as well as for the others who use and apply surveying measurements such as civil engineers, architects, and attorneys. It is intended to explain surveying *measurement* more so than surveying techniques and methods. Since the world of the surveyor is principally one of measurement, a truly competent surveying technician or professional surveyor must understand the nature of measurement and not simply the techniques, methods, and computations. This point will be further developed throughout the book and discussed in Chapter 9.

1-2 THE GENERAL NATURE OF MEASUREMENT

Misunderstandings about measurements often exist in the minds of laymen and associated professionals with whom surveyors work, as well as in the minds of many practicing surveyors themselves. The misunderstanding, it can be theorized, might be traced to the manner in which most were taught arithmetic. We were taught to *count*, not measure. We learned that two plus two is four and it went unquestioned. Decimals were introduced but often in the context of money (again, *counts*) or in association with fractions (again, *exact* divisions). We were never taught that anything *measured* — weight, volume, distance, angle, area, height, time — cannot be treated the same as a count. Many people, for example, accept their height or weight to be exact numbers without question. But, when one visits a doctor for a physical examination, his or her weight is rounded to some significant digit according to how close the scale can be read and how close the nurse chooses to read it. So, it is inexact simply because of reading uncertainties. It may also be even more incorrect if the scale is not properly calibrated!

If the number of people in a room is counted, the result is an exact number (integer) and there is no decimal. Likewise, in counting money, the value found is an exact number. It may have a decimal when written as dollars, and look like a measured value, but it is very much different. The number of cents is an integer, thus the total is still a count, not a measurement. If one has $32.46, there is no error. The person has exactly 3,246 pennies, no more and no less, unless a mistake in counting has occurred. But if a distance has been read as 32.46 feet, the true value may actually be 32.458216. . . , 32.464512. . . , or even closer to 32.50, depending on the calibration of the equipment, refinement in reading, general care in measuring, and a number of other phenomena. Beginning surveying students should attempt to accept the fact that measurements are inexact. But, accepting this fact is not an excuse for sloppy work, because a level of accuracy should be achieved which is consistent with that needed according to specifications and as dictated by economy and available equipment. The evaluation and analysis of the inexactness (more properly called *error*) will be studied further in Chapter 2.

Survey measurements are often said to be made in the "three elements of space." This means that surveyors are involved with measuring three dimensional space rather than just on plane surfaces. The three dimensions are generally given names such as *distance, elevation,* and *direction*. Unless otherwise described, "distance" in surveying is construed to be *horizontal distance*. It is the component which lies perpendicular to vertical. Therefore, a distance along a slope (incline) must be reduced to horizontal. Making computations on a plane surface for limited areas (or on a spheroid for geodetic surveying) can be done properly only if the horizontal component of a slope distance is used. "Elevation" is vertical distance measured above reference surfaces called *datums*. "Direction" is an angular measurement from a meridian (north-south) line to a particular line on the surface of the earth. North, southeast, and northwest, are general directions.

Thus, surveying measurements are usually either linear or angular, although time, barometric pressure, humidity, gravity, stream flow rate, area, volume, weight or tension (mass), temperature, slope, reflected or emitted electromagnetic and other energy, are also measured by surveyors. Many such measurements, such as slope, area, and volume are not made directly but are computed from distances and angles.

Most surveying measurements are made on or near the surface of the earth and involve measurements of a magnitude much larger than the familiar ones made in a classroom, home, or office, using 12-inch rulers and protractors. Laboratory measurements, such as those made from maps or photographs or for plotting maps or other scaled reproductions of terrain, must usually be made with high precision. Although devices similar in principle to rulers and protractors are often employed a considerable amount of technical operator skill and equipment design is necessary to achieve the precision needed.

Surveyors and surveying technicians, either in the field or laboratory, must understand the nature of measurement and its relationship to the instrumentation and precision requirements of any job.

1-3 APPLICATION OF SURVEYING MEASUREMENTS

Surveying measurements are usually required whenever and wherever land is divided or new construction is involved. Surveying measurement is a part of all engineering design and construction. Profiles, cross-sections, and maps showing terrain and drainage features must be plotted for large projects. All projects, large or small, require a certain amount of field measurement to set stakes and other control points to guide construction. Land development projects require that measurement in all three elements of space be made before, during, and after development. The expertise of the surveying professional is required by the engineer, as well as by other professionals such as architects and landscape architects.

Whenever land is divided or there is a need to delineate established boundaries of real property, measurements are required. Establishing property boundaries has historically been one of the primary functions of surveyors. Understanding the nature of measurement is an absolute necessity in order to properly locate and preserve the position of property lines.

Some form of surveying measurements are employed in map making, mining, forestry, geology, industrial operations, natural resources studies, land use planning, space exploration, observational astronomy, and development of modern cadastral systems toward complete inventory and knowledge of the land as a resource and a commodity.

1-4 CLASSIFICATION OF SURVEYING

Surveying can be classified in several ways. Classification by function or scope helps to reveal specialties within the practice and to further explain or define the profession. It should be understood that a well rounded professional surveyor would have a good working knowledge of all of the classifications though he or she may be a specialist in only one or a few of them. An engineer should at least be able to identify the categories in order to utilize surveying measurements properly. The engineer should understand the aspects of surveying practice which relate to engineering and mapping in particular.

The simplest classification of surveying would place all functions into two groups: (1) plane surveying, and (2) geodetic surveying. *Plane surveying* assumes that the earth is a plane surface and that north-south lines are all parallel. It should be obvious that this method of surveying is suitable only for limited areas since the earth is not flat and north-south lines actually converge and meet at the poles. But, calculations on a plane are relatively simple. Plane geometry and plane trigonometry are used. Plane surveying calculations can usually be employed for areas of a few square miles before too much error is caused by neglecting curvature of the earth and convergence of the meridians. The exact limits are governed by how much error from these sources is permissible for a specific surveying project. *Geodetic Surveying* considers the earth's curvature and meridian convergence. Although the measurement techniques and equipment used in plane and geodetic surveying are nearly the same, the calculations employed in geodetic surveying involve treating the earth as a spheroid or ellipsoid instead of a plane. Geodetic surveys are generally made over large areas, such as states or continents, and the specifications for measurement tolerances are quite strict.

Most of the remainder of the classifications of surveying to be discussed in this section employ plane, rather than geodetic surveying. Likewise, much of the work of the engineer and practicing surveyor involves plane surveying. After recognizing the difference between plane and geodetic surveying, the science and art can be further classified into the following categories.

Land Surveying. This classification is often called property surveying or cadastral surveying. In the United States, as in most countries which have licensing requirements, each state (or province, etc.) has a legal definition as a part of registration or licensing laws. Usually, the definition in the U.S. relates primarily to the location or retracement of land boundaries, property lines, or property corners which define real property. In most states, this classification also includes the preparation of the plats and land descriptions associated with the property surveys, and also the determination of the shape or contour of the land and preparation of maps. It may also include new land subdivision design and layout, as well as special functions such as mine surveys. In other developed countries, it may also include town planning and other functions associated with land.

Although the work of land surveyors is usually performed "on the plane," such surveyors often use geodetic methods when their work requires establishing control over large areas. The land surveyor has traditionally been involved primarily with property line surveys, but this class of surveyor has started to become the "general practitioner" of the surveying profession and may have knowledge in all of the specialties. Indeed, proper and thorough analysis and execution of most land surveys necessitates the application of many modern methods and theoretical concepts in order to properly perpetuate survey evidence for the benefit of the public and future generations of surveyors. As educational opportunities become wider, broadening of the function of a land surveyor will probably continue.

Topographic Surveying. This classification has been defined as surveying "which has for its major purposes the determination of the configuration (relief) of the surface of the earth (ground) and the location of natural and artificial objects thereon." (Reference 4) The word topography itself means "the features of the actual surface of the earth considered collectively as to form." (Reference 4) A single feature such as a mountain is a topographic feature. Topography is subdivided into *hypsography* (the relief features), *hydrography* (the water and drainage features), and *culture* (man-made features).

Topographic surveys were formerly made entirely by using ground methods such as transit-stadia or plane table. Nowadays, aerial photogrammetry is used almost exclusively for topographic surveys of large areas, although the older ground methods are still often feasible for small areas.

Hydrographic Surveying. Although this term was mentioned as a part of topographic surveying, it is often considered as a separate classification. Hydrographic surveys are surveys of lakes, streams, reservoirs, oceans, and other bodies of water. They may consist of determining elevations of shorelines and the bottom or bed of the water body, stream flow rates, or the capacity of reservoirs. Special purpose equipment is required for many of these surveys.

Marine Surveys. Although related to hydrographic surveying, this term is often considered in a broader sense. Marine surveys encompass aspects of position determination for platforms or moving vehicles at sea, and surveys for hydrography, nautical and bathymetric mapping or charting. The subject includes the science and technology of navigation, hydrography, bathymetric and control surveying, the theory of tides and dynamic motions and shape of the oceans and their observations.

Engineering Surveying. This classification includes all surveying related to design and layout of engineering works. Civil engineering is the specific branch of engineering which most commonly employs surveying measurement.

There are various sub-classifications of engineering surveying. One is *route surveying* which involves the alignment of engineering works associated with transportation and communication systems such as highways, railways, canals, pipelines, and transmission lines. Route surveys are linear, or "point-to-point" in nature. Route surveys include profiles, cross-sections, and sometimes complete topographic mapping of narrow linear strips of land along the proposed route alignment. Such surveys are made prior to route design. After design, there is another series of surveys required which may involve the layout of either straight lines or circular, spiral, or parabolic curves. This is required prior to construction of the route and is a part of the type of surveying discussed next.

Another subclassification of engineering surveying is *construction surveying*. After engineers, architects, surveyors, or planners design or plan any route, building, or other structure, there must be grade stakes and other control stations set to guide construction. Surveyors interpret plans and coordinate with engineers and contractors for the layout and construction staking. Much ingenuity is involved with complicated structures, the control network can become intricate, and the accuracy tolerances are often strict. Since millions of construction dollars are often involved, the work must be very systematic so as to avoid mistakes in establishing the control or misinterpretation by engineers or contractors. Civil engineers should be especially aware of the complexities of the art of construction surveying. Planning the method of staking construction, as well as execution of the measurements, often requires judgment possessed only by a professional surveyor who thoroughly understands the nature of measurement as well as the construction process.

In a sense, many large-scale topographic surveys are engineering surveys since they are used for plotting maps from which engineering designs are made. Engineering surveys can be considered as falling into two groups: (1) those made prior to design (topographic surveys, profiles, cross-sections, etc.) and (2) those made prior to construction (grade stakes, alignment stakes, other layout control points). A third group called "as-built" surveys could also fall under the heading of engineering surveys. These are surveys to determine positions of structures as actually constructed. These are often required to check the contractor's work, assure that the structures will function according to design, and provide a record of locations of structures for maintenance and other subsequent design purposes.

Other classifications. Reference 5 at the end of this chapter gives definitions for other classifications of surveying, such as exploratory, geologic, industrial, inventory, location, mining, photogrammetric, and soil surveying. These are, in a more exact sense, not classifications but special types of surveying. *Photogrammetry* or photogrammetric surveying, for example, is a measurement method or tool employed in most of the classifications of surveys discussed, particularly topographic surveying, and is not actually a separate classification. *Remote sensing* is a special type of environmental surveying, usually done from airplanes or satellites, where precise measurements are made of reflected or radiated electromagnetic or thermal energy which is translated or transformed into an image or other form for further study. All such specialties rightfully fall under the broad heading of surveying since they employ measurements of the land for engineering, planning, or other purposes. It is felt that the broad classifications discussed form the major categories of surveying practice.

1-5 UNITS OF MEASUREMENT IN SURVEYING

Linear Units

Survey distance measurements have been made throughout history using a variety of units. Ancient units, for example, included the cubit, palm, and digit. Units such as the *rod* were employed in Medieval England and later in the United States and in other countries. The rod has also been called perch, rood, pole, and other names. Even later, the *chain* was defined as a unit of measurement, following the invention of a Gunter's chain by Edmund Gunter in 1653. Rods and chains were used for early land surveys and such units are still seen in old land surveys and descriptions.

The English foot is still employed in surveying, particularly in engineering measurements. In order to facilitate computations, surveyors divide the foot into decimal units (tenths, hundredths, etc.) rather than inches and fractions of inches.

Surveyors and engineers are rapidly becoming more comfortable with metric units. Students should learn to think in this system since conversion to metric units is progressing rapidly.

In certain parts of the U.S., a unit of Spanish origin called the *vara* has been used in land surveys. Surveyors should be aware of the typical units of linear measurement and be able to make conversions. Some values which may be unfamiliar are as follows:

1 rod = 16.5 feet

1 Gunter's chain = 66 feet = 100 links = 4 rods

1 mile = 80 Gunter's chains

1 nautical mile = 6076.10 feet

1 Engineer's chain = 100 feet

1 Meter = 39.37 inches (approximate)

1 Vara = 33 inches in California = 33⅓ inches in Texas

Some particulars should be explained about metric conversions lest the unwary student become trapped in the confusion. In 1791, the meter was introduced as the unit of linear measurement in the metric system, being defined as 10^{-7} of the earth's quadrant. The definition was changed, in practice, in 1799 when it was defined by the length of a prototype meter bar. The meter was redefined again in 1960 by the 11th General Conference of Weights and Measures in terms of the wavelength λ_k, of the radiation produced by a specified quantum transition of the Krypton-86 isotope. This permits a more accurate reproduction of the meter than is possible through use of a bar.

The SI (metric system) unit of length defined by the General Conference is

$$1 \text{ meter} = 1{,}650{,}763.73 \quad \lambda_k.$$

All measuring devices such as wires, tapes, and rods may be calibrated in terms of the meter, so that all distances can be measured uniformly in the SI system. In 1866, the U.S. Congress legalized the use of the metric system and adopted the Paris International prototype meter. The meter was then equivalent to 39.37 inches exactly which yields,

1	Foot	=	0.3048006096 meters,
1	Meter	=	3.28083333 feet,
1	Inch	=	2.540005080 centimeters.

But in 1960, when the foot was redefined, the result was

1	Foot	=	0.3048 meters exactly,
1	Meter	=	3.280839895 feet,
1	Inch	=	2.54 centimeters exactly.

However, the surveying profession in the U.S. continued to use the "old" (prior to 1960) conversion because the national geodetic network of horizontal control is based on this. Until there is a readjustment of this control system, surveyors will continue to use the values established prior to 1960. This readjustment is to be completed in 1983. Therefore, the "survey foot" equivalent will change from 1 meter = 3.280833 feet to 1 meter = 3.280840 feet in 1983.

Precise measurements from maps, photographs, stereoscopic photogrammetric models, and other imagery are made using units such as microns or millimeters. Conversions of these, and other units in the metric system are as follows:

1 millimeter	=	1,000 microns (micrometers)
1 centimeter	=	10 millimeters
1 meter	=	100 centimeters
1 kilometer	=	1,000 meters

Angular Units

The angular system used in surveying in the United States and many other countries is the sexagesimal system in which the circle is divided into 360 degrees which in turn are divided into 60 arc-minutes which are further divided into 60 arc-seconds. In some other parts of the world, particularly Europe, a centesimal system is used in which the circle is divided into 400 parts called grads (sometimes spelled grades) which is in turn divided into 100 centigrads or centesimal minutes which are further divided into 100 decimilligrads or centesimal seconds. As can be seen, this latter system has the advantage of using decimal numbers.

Surveyors often find the radian to be a useful angular unit when making calculations employing small angles and when considering measurement error analysis involving angles. A radian is the angle subtended at the center of a circle by the arc equal in length to the radius of the circle. It is therefore equal to $360° \div 2\pi$ or approximately 57° 17′ 44.8″. Conversions between radians and sexagesimal and centesimal units are, to eight decimal places,

$$
\begin{aligned}
1 \text{ Radian}, \rho &= 57.295778° \\
&= 3437.7467' \\
&= 206{,}264.80''
\end{aligned}
\qquad
\begin{aligned}
1 \text{ Radian}, \rho &= 63.661977^{g} \\
&= 6366.1977^{c} \\
&= 636{,}619.77^{cc}
\end{aligned}
$$

Surveyors find it useful to remember that 1″ of arc is 1 ÷ 206,265 or approximately 0.0000048485 radians which is also the sine or tangent of 1″, and that 1′ of arc is 0.00029089 radians which is also the sine or tangent of 1′ to 5 significant figures. A rough rule of thumb in this regard is that 0.3 units subtend 1′ of arc for 1,000 units of distance. The above conversions are easily made on modern hand-held scientific calculators having the pi function.

Area Units

At present, the most commonly used land unit of area in the United States is the *acre* which derives from an ancient Roman term called the "ager", meaning the amount of land a yoke of oxen could plow in one day. The acre was more precisely defined in Medieval England as an area 4 × 40 poles or rods. When Edmund Gunter invented his chain in 1653, he deliberately made it 4 rods long in order to easily lay out an acre. It can be seen that an acre is 1 × 10 chains or 10 square chains and one has only to move the decimal to the left one place to arrive at the acreage. Since one chain equals 66 feet, an acre is 43,560 square feet.

The metric area unit is the *hectare* which is 10,000 square meters. It can be shown that this is approximately 107,639 square feet or 2.47105 acres. Some surveyors are attempting to implement the use of the hectare as an alternate or preferred unit since distance units are rapidly becoming accepted in the metric system. The future of the acceptance of the hectare and other metric units among surveyors and allied professionals is unknown at this time.

1-6 MISCELLANEOUS CONCEPTS AND DEFINITIONS

The Size and Shape of the Earth

Since most surveying measurements are made on or near the surface of the earth, a general knowledge of the size and shape of this reference surface, and the various reference systems associated with the earth should be understood.

The earth's shape approximates an oblate spheroid of revolution, the length of the equatorial axis being approximately 27 miles greater than the polar axis. Relative to the size of the earth, this difference in diameter is very small and an average radius of the earth can be used for all but the most precise geodetic calculations. The spheroid used in the United States is called Clarke Spheroid of 1866. This spheroid has a polar radius of 20,855,121 feet according to some references and 20,854,892 feet according to other references. The equatorial radius is given as either 20,926,062 feet or 20,925,832 feet in different references. The average value commonly used for the radius in the United States is 20,906,000 feet. There are several other spheroids used in geodesy throughout the world.

The spheroid mentioned above is a regular, smooth surface at *mean sea level,* an average height of the sea at all stages of the tide. Actually, mean sea level is simply the average level of the sea at a specified place. In 1929, in the United States and Canada, the average level of the sea was taken from 26 tide stations using records covering a 19 year time period. This average was held fixed and became known as the "Sea Level Datum of 1929". A *datum* is a reference surface. The name of the datum was changed in 1973 to "National Geodetic Vertical Datum of 1929." Even though actual sea level averages vary, the average of 1929 is always used in the interest of maintaining a permanent value for engineering and other work.

Actually, the mean sea level reference surface is not perfectly smooth because of variations in the direction of gravity caused by various topographical features. The true surface, which is everywhere perpendicular to the plumb line or direction of gravity, is called the *geoid.* The best fitting mathematical spheroid will approximate this surface within about 100 meters at any point. For most survey work, we can assume that the direction of gravity is perpendicular to the spheroid, neglecting the local gravity effects.

A *level surface* is a surface every element of which is normal to the plumb line. The sea level datum is a level surface, but there are an infinite number of level surfaces above and below sea level, all approximately parallel with it.

If a plane is passed through the earth's center, its intersection with a level surface forms a *great circle*. Any portion of such a circle is called a *level line*. If a plane is passed through the poles as well as the earth's center, the line along any level surface is called a *meridian* or a line of longitude.

If three points are located on the earth and great circles are passed through each two adjacent points, a *spherical triangle* is formed. It should be obvious that the spherical angles between these great circles at each point are greater than the plane angles between straight lines connecting the points. The difference between the sum of the spherical angles and the plane angles is called *spherical excess*. For a triangle of about 75.5 square miles area, this difference is only one second of arc. This gives some idea of the effect of the earth's curvature and when it must be considered. It is also useful to mention that an arc 11.5 miles long lying in the earth's surface, differs in length from the subtended chord by only 0.05 feet. From this we can appreciate why plane surveying techniques can be employed for most survey work. See Figures 1.1 - 1.3.

Surveying Terms

Elevations are vertical distances above a chosen datum. Thus, each point on the earth's surface has an elevation above sea level or any arbitrarily chosen

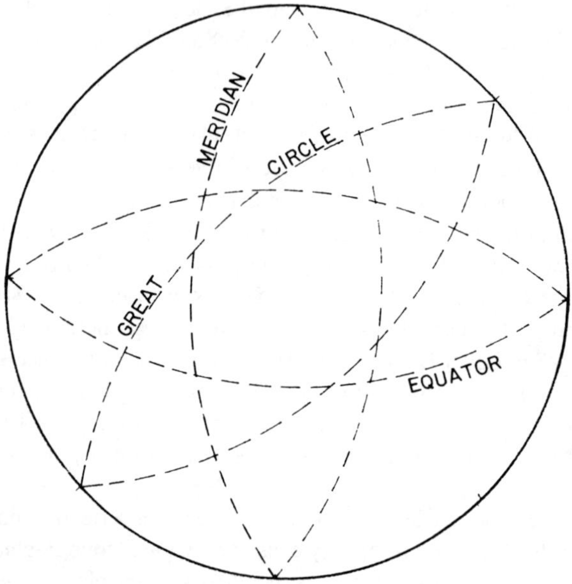

Figure 1.1 Great Circles and Meridians

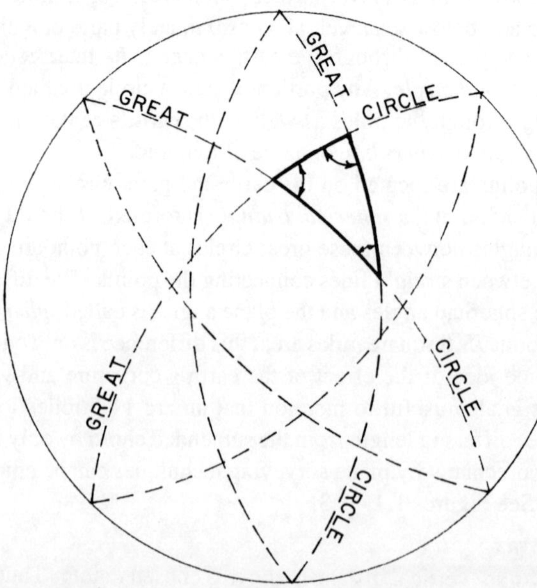

Figure 1.2 Spherical Triangle

datum. Elevations are measured vertically or along plumb lines. *Elevation differences* are vertical distances between level surfaces or level lines.

A *vertical line* follows the direction of gravity at a point. It is perpendicular to the plane of the horizon. Vertical lines through several different ground points do not intersect at one point. Surveying students eventually gain appreciation for the

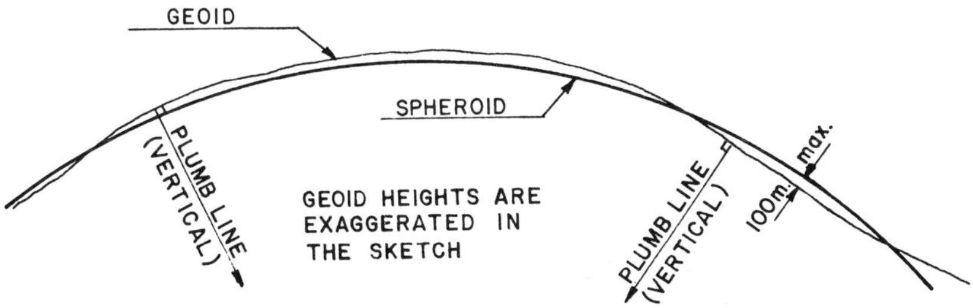

Figure 1.3 Level Surface and Geoid

fact that gravity is the primary reference to which all lines and surfaces are referenced in surveying. A *vertical plane* is a plane that has a vertical line as an element. A *horizontal line* at a point is a line perpendicular to the vertical line at that point or one which is tangent to a level surface. Likewise, a *horizontal plane* is a plane perpendicular to a vertical line at a particular point or a plane which is tangent to a level surface. See Figure 1.4.

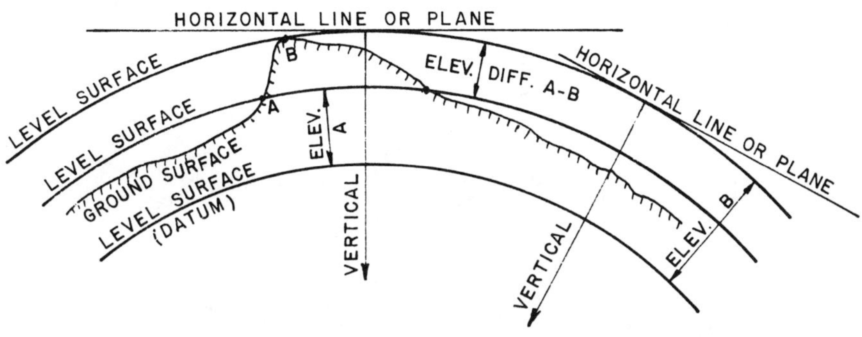

Figure 1.4 Elevations and Vertical Lines

A *horizontal angle* is an angle measured in a horizontal plane between two vertical planes. A *vertical angle* is an angle measured in a vertical plane and, in surveying, is measured from a horizontal line or plane. Vertical angles are either angles of *inclination* (positive, measured upward from horizontal) or *depression* (negative, measured downward from horizontal). Positive angles are also called *elevation* angles (also called *altitude* in astronomy). A *zenith angle* (also called zenith distance in astronomy) is another variation of vertical angle, the difference being that zenith angles are measured down from the upward direction of the

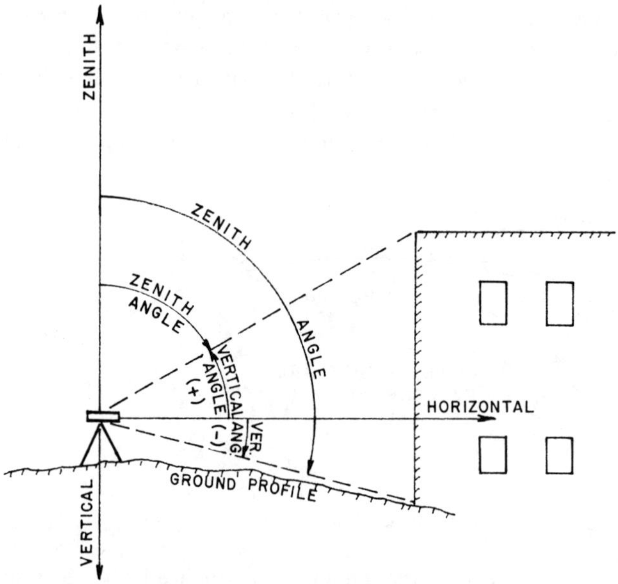

Figure 1.5 Zenith Angle and Vertical Angle

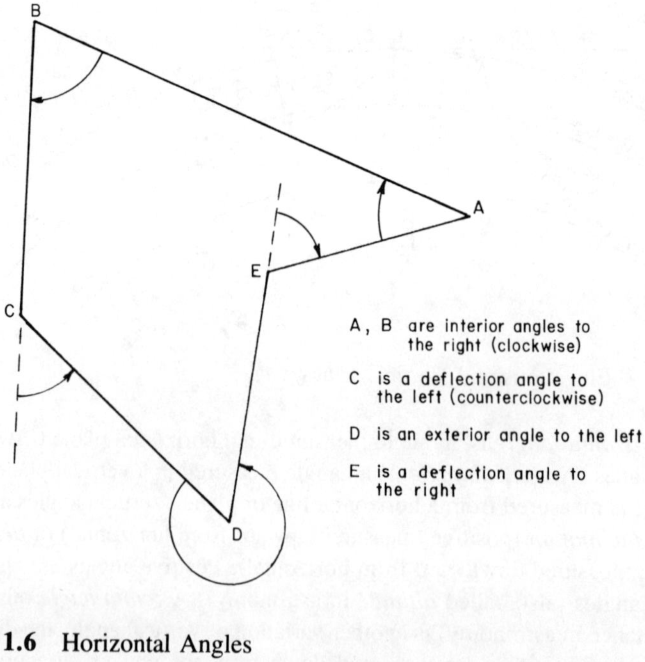

Figure 1.6 Horizontal Angles

plumb line. Thus, a zenith angle of less than 90° would be a positive vertical angle and one greater than 90° would be a negative vertical angle. Many modern angle measuring instruments read zenith angles instead of vertical angles. American vernier transits generally read vertical angles. See Figures 1.5 - 1.6.

Direction of a survey line is the horizontal angle from a vertical plane through a reference line to the vertical plane through the survey line. Reference lines can be arbitrarily chosen but more often they are meridian lines. Directions of survey lines are usually referenced to the north direction. It is not the purpose in this book to explain all of the various differences in this north direction — suffice to say that there are differences between and among magnetic, geodetic, astronomic, grid, and mere assumptions for north. The determination of a precise reference direction is left as subject matter for other texts. There are two methods of stating direction with respect to a reference meridian. The *azimuth* of a survey line is the horizontal clockwise angle from a meridian to the survey line (generally from the north). Azimuths vary from 0° to 360°. The *bearing* of a survey line is an acute angle measured either clockwise or counterclockwise from a meridian line. A statement of bearing is incomplete without the quadrant designation. Quadrants are either northeast, southeast, southwest, or northwest. In surveying, these are often called the first, second, third, and fourth quadrants, respectively. Since bearing angles are confined to quadrants, they vary from 0° to 90°. For example, there would be no such thing as a bearing of N 91° E. If a line was 1° toward the south direction from due east, it would be S 89° E. It is noted that this is also an azimuth of 91°. Note also that bearings are designated by stating either north or south first, then the angle, then either east or west. See Fig. 1.7a and 1.7b.

Distance is usually construed to be *horizontal distance* which is the distance between the two points projected onto a horizontal plane. Horizontal distance can also be defined as the distance between the two points measured along a level line. Practically speaking, these definitions are the same, since reduction to horizontal has meaning only in plane surveying. In geodetic surveying, distances are reduced to those along level lines and usually to sea level distances, and slope corrections are made in the process.

The terms *slope* and *grade* have similar meaning in surveying. In one sense they are synonymous and mean the vertical divided by horizontal distance, multiplied by 100. For example, a 2-foot ground elevation difference over a distance of 100 feet would represent an average 2% ground *slope* or *grade*. Slope can also be represented by a vertical angle. Thus, it could be shown that a 2% slope is a slope angle (vertical angle) of about 1° 08′ 45″. The word "grade" sometimes also means "elevation" and is generally used when referring to the grade or elevation of something being constructed, such as a pavement or foundation.

Many other terms will be defined as the various topics are developed in this book.

1-7 HISTORICAL ASPECTS OF SURVEYING

Ancient-Medieval Surveying

Surveying and measurement is as old as recorded history itself. The art and science of surveying has evolved out of an ancient need to establish positions on

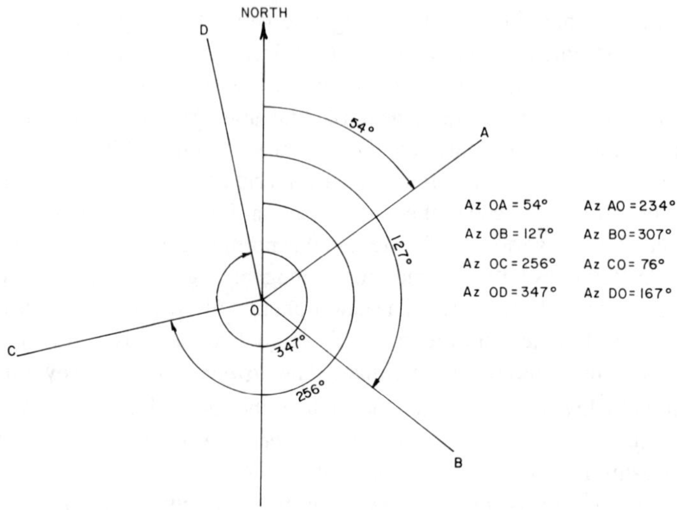

Figure 1.7a Azimuths

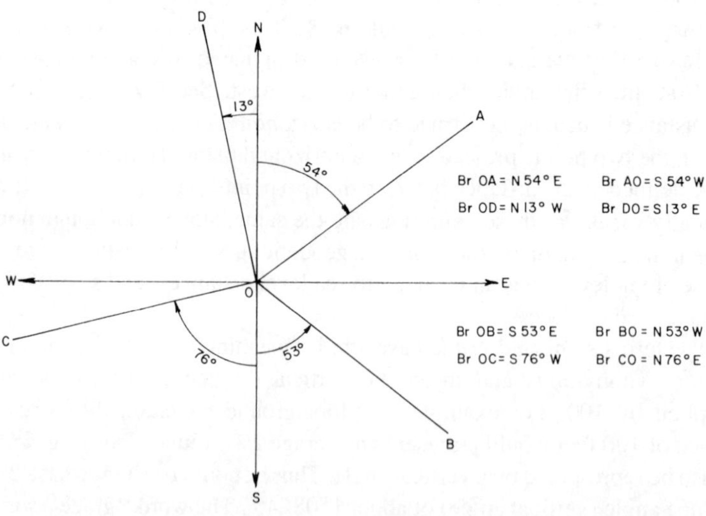

Figure 1.7b Bearings

the surface of the earth. Mathematical theorems, such as the famed Pythagorean theorem were developed from such a need. Geometrical principles and surveying have developed together. The sophistication of early instruments was not high. Horizontal distances and horizontal right angles were about all that were measured in surveying in ancient times. Vertical angles were sometimes measured, but mostly in astronomical observations. Acute horizontal angle measurements were rarely made. Elevation measurements were mostly confined to determining a horizontal line — differential leveling and slope measurements, in the modern sense were not usually made. Surveying computations were avoided whenever possible. The science of geometry was just developing in ancient times and tools like logarithms and slide rules were not invented until centuries later. The use of gravity as a reference, using plumb bobs and sometimes water to establish vertical and horizontal alignment, has been universally employed throughout the ages. See Figs. 1.8 - 1.10.

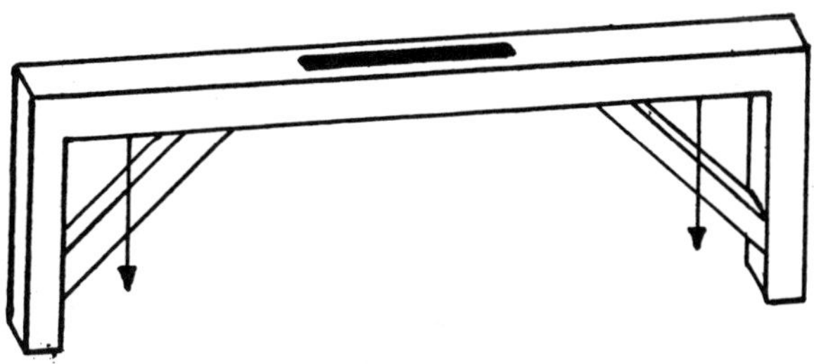

Figure 1.8 The Chorobates (Courtesy of CARBEN Surveying Reprints)

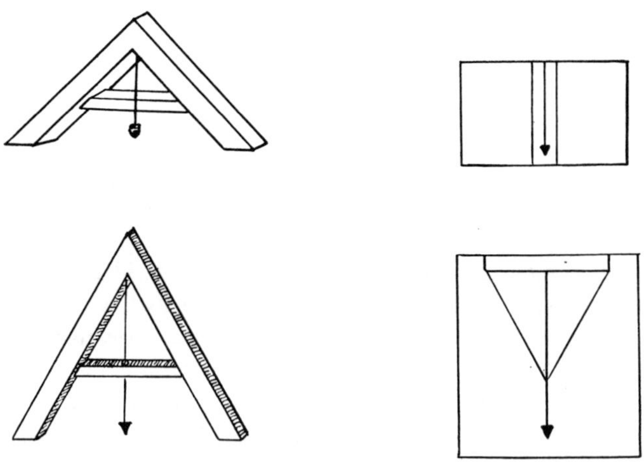

Figure 1.9 Plumb bob Levels (Courtesy of CARBEN Surveying Reprints)

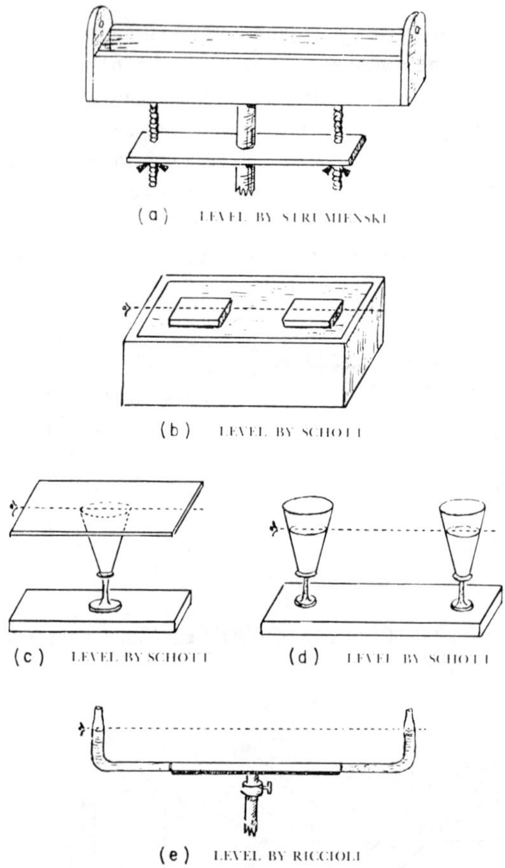

Figure 1.10 Water Levels (Courtesy of CARBEN Surveying Reprints)

Early developments in astronomy, navigation, and artillery directly affected developments in surveying, although surveyors did not always immediately apply the same instruments. Gradually, surveying instrumentation evolved. Instruments utilizing the same principles as some originally developed in medieval times and the early Renaissance, are still used today. Only recently have technological breakthroughs drastically changed the basic tools. The function of at least one ancient "universal" instrument is the same as that of modern ones. See Figure 1.11.

Surveying practice has evolved from a need to locate, describe, and mark something on the earth's surface. When there was no civilization, or in areas where there are still no concepts of land possession and ownership, there is little need for surveying. Nomadic tribes and prehistoric man had no need for surveying. Our original Americans, the Indians, had no need for measurements since their idea of land ownership was that it all belonged to the tribe anyway. When any primitive society begins to become civilized, build water supply and irrigation systems, roads and bridges, sewage and drainage systems, and other structures, then

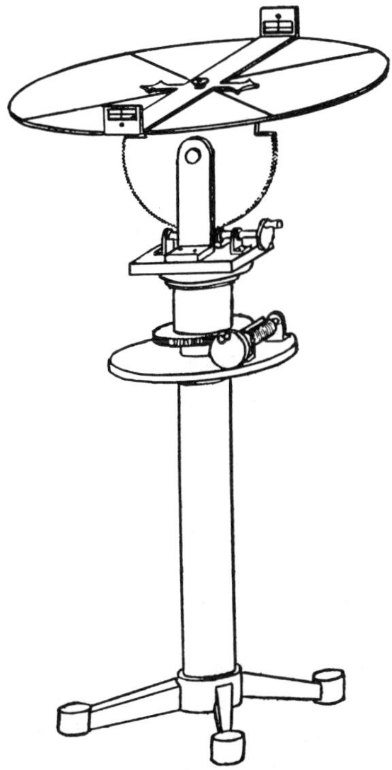

Figure 1.11 The Dioptra (Courtesy of CARBEN Surveying Reprints)

measurements in the three elements of space are needed. When man begins to reach a level of settled agricultural development with population large enough to compete for land, then land division is needed, along with possible recording of such divisions, assuming individual or collective ownership of some form is recognized.

Throughout history, empires and private land claims have been either lost or created as a result of armed or legal conflicts arising over the challenge of some claim to land. Directly or indirectly, surveying has been associated with these changes, and even in ancient times, both the legal and professional aspects of surveying were recognized.

Renaissance Surveying

There were several influences on the development of surveying instruments and practices during the Renaissance. An increased interest in navigation, astronomy, and cartography led to development of surveying related instruments. The change in attitudes about land ownership created a need for surveyors to lay out boundary lines. The art of printing developed and this led to the spread of knowledge of all kinds.

In Europe, these various influences caused a movement toward the development of scientific instruments and created a need for surveyors. Several books were written for surveyors in the sixteenth century and many people started improving old versions of instruments and developing new ones. Those who were directly involved with the development of instruments have been called ''mathematical practioners.'' This movement continued in Europe through the following two centuries.

During the first century after Columbus' discovery, we can assume that little was done in the New World related to surveying, except for some exploratory surveys. Settlements were eventually established, but geographic delineations were poor and there was no desire for accurate surveys. As a number of people began to immigrate and establish settlements, a need for surveying was created. Inland waterways had to be investigated as potential transportation routes, the topography had to be learned in order to avoid certain dangers, and boundaries had to be delineated. As the early settlements grew into colonies, all of the surveying and mapping needs were multiplied. In answer to the need for surveying, a phenomenon similar to that in England eventually occurred. Several people started to make instruments for colonial surveying needs. Most were patterned after existing ones, being primarily compasses and plane tables. There was one difference between these practitioners and those of England. Whereas, the quality of work in England was fairly carefully controlled and regulated, the American instrument makers only occasionally had any specialized training and they were subject to no regulation or controls on their work quality. They, nevertheless, produced some equipment that eventually was of a quality comparable to some of the English instruments. It must be assumed, however, that many of the initial surveying, navigation, and exploratory instruments were brought from England. Naturally, most of the scientific knowledge originated there also, and some surveyors immigrated to the New World. Fig. 1.12 is a Renaissance mine surveying compass.

Evolution of Modern Surveying

After the Revolution in the United States, surveying activity increased. Vast areas, including the Northwest Territory, the Louisiana Purchase, and eventually the western states reaching to the Pacific Ocean were surveyed and marked according to the U.S. Rectangular System of Surveys. Maps were prepared in large numbers, at first for military purposes and eventually for general civilian use. The layout of new towns and cities required surveying. Canal and railroad construction demanded the development of new techniques and instruments. The science of geodesy was expanding and geodetic surveying began to employ new, precise instruments. Geodetic control was first established along the east coast of the U.S. in the early nineteenth century. Such control had been existent in Europe decades prior to the application in this country.

The twentieth century, and particularly the decades of the sixties and seventies, has seen a revolution in the instrumentation employed in surveying. Electronic computers, calculators, and measuring equipment have changed surveying practice immensely. Also photogrammetric techniques have caused significant changes in measurements for maps and other uses. Higher land values have created a need for better instruments and more accurate locations.

Figure 1.12 Renaissance Compass (Courtesy of Crown Copyright. Science Museum, London.)

There has always been a need for surveying and there always will be. What the future holds cannot be determined. We can only assume that equipment will become more refined and automated, that technicians will be needed to make many measurements of land and the environment, and that professionals will be needed to interpret and analyze the measurements. More will be discussed on the future of surveying measurements in Section 9-7.

Problems - Chapter 1

1.1 With what two fundamental problems in measurement are surveyors faced?

1.2 What is the basic difference between a surveying technician and a professional surveyor?

1.3 Explain the difference between a count of 12 inches to a foot and a measurement of 12.0 inches.

1.4 What are the two most common types of measurements made by surveyors?

1.5 What are the two simplest classifications of surveying?

1.6 What is the difference between land and engineering surveying?

1.7 Distinguish between hydrographic and marine surveying.

1.8 Convert the following to feet:

14 rods

40 Gunter's chains

103.5 meters

1.9 Convert the following:

0° 01′ 30″ to radians

0.0000097 radians to seconds of arc

0.05 radians to grads

1.10 What would be the angle subtended between the left and right sides of a man's thumb at a distance of 40 meters if the thumb width is 2.5cm?

1.11 How many acres are there in a rectangular tract of land 2 rods by 20 rods?

1.12 Why is the average radius value used in the United States less than the equatorial radius of the earth?

1.13 Define the following:

a) datum
b) geoid
c) meridian
d) vertical plane
e) horizontal angle
f) zenith angle
g) azimuth

1.14 Explain why the sum of the interior angles of a spherical triangle is greater than the sum of the angles of a plane triangle.

1.15 Convert a vertical angle of −14° 16′ to zenith angle.

1.16 Convert the following:

Bearing N 16° 18′ W to azimuth

Azimuth 175° 16′ 20″ to bearing

1.17 Convert a slope of −2.5% to:

a) vertical angle b) zenith angle

1.18 How could a carpenter construct a crude construction level using two paper cups?

1.19 How would interest in navigation and cartography influence development of surveying methods?

1.20 What two things related to civilized man creates a continuing need for surveying practice?

Additional Reading References - Chapter 1

1. Barry, Austin B. *Errors in Practical Measurement in Science, Engineering, and Technology.* New York: John Wiley & Sons, Inc. 1978, Chapter 1.

2. Bedini, Silvio A. *Thinkers and Tinkers.* New York: Charles Scribner's Sons, 1975.

3. Blachut, T.J., Chrzanowski, A., and Saastamoinen, J.H., *Urban Surveying and Mapping.* New York: Springer-Verlag, 1979, Chapter 3 (Introduction).

4. Brinker, Russell C. and Wolf, Paul R. *Elementary Surveying.* Sixth Ed., New York: Harper and Row, 1977, Chapters 1,6.

5. *Definitions of Surveying and Associated Terms.* American Congress on Surveying and Mapping, Falls Church, Virginia, 1978.

6. Kiely, Edmund R. *Surveying Instruments: Their History.* Columbus, Ohio: CARBEN Surveying Reprints, 1979.

7. Schmidt, Milton O. and Rayner, William H. *Fundamentals of Surveying.* Second Ed. New York: D. Van Nostrand Co., 1978, Chapter 1.

CHAPTER 2

THE WORLD OF MEASUREMENT

2-1 THE INEXACTNESS OF MEASUREMENT

Anything measured is inexact. That is, the exact or true value, though it may exist, can never be determined by humans or machines, no matter how refined the techniques are. It may be possible to measure with a high degree of accuracy, obtaining values significant to several decimal places, but the true value may contain an infinite number of digits which would be impossible to determine. But, this does not mean that a high degree of refinement in measurement is impossible. It simply means that *exactness* is not possible.

A popular demonstration often used to show inexactness is to compare measurements made of the width of a desk or table by several people. From outward appearances, the desk is rectangular and everybody's scale is the same; therefore, logic would dictate that all should get the same value when measuring it. But, then why did 10 intelligent and bright-eyed engineering students, all using 12-inch engineer's scales, measure the same table width and observe different values as follows?

48.05 inches	47.86 inches
47.94	47.94
47.92	47.97
47.96	47.90
47.94	47.89

One could reason that the table edge varied a little from a straight line, each person aligned the scale at a slightly different spot on each side of the table, the 12-inch lengths were not laid out in a perfectly straight line, making marks on the table for each 12-inch length was done with varying care, the measurements varied from being exactly perpendicular to the table edge and the amount of non-perpendicularity varied with each individual, estimation of the final reading caused variations, and there were variations in skill and care of each person. This example demonstrates not only the inexactness of measurements, but some of the reasons for inexactness–namely people, equipment, and nature itself.

2-2 ERRORS AND MISTAKES

The term *error* means the difference between an observed or computed value of a quantity and the true value of that quantity. The *true value* is the actual value existing in nature which is seldom known except when fixed by mathematics (i.e., the sum of the angles of a triangle = 180°) or by authority (i.e., international prototype meter). The *observed value* is generally called the *reading* in surveying. It is what is read directly or accumulated on the scales of the equipment. From the above definition, the following simple equation can be written:

$$\text{Error} = \text{Reading} - \text{True Value} \tag{2.1}$$

Errors are not to be confused with *mistakes*, the latter being blunders caused by the observer (false reading) or by a defective instrument, or both. Mistakes can be traced to inattention, carelessness, improper training, bad habits, or lack of innate ability. Equipment defects can be traced to these personal faults since equipment should be properly checked before it is used. Beginning surveying students should recognize that mistakes are not in the same category as errors. Most readers probably have been taught in everyday terminology that errors and mistakes are synonymous. This thinking habit must be immediately altered in order to begin to learn the rudiments of survey measurement theory. Mistakes can be avoided and most errors cannot, as will be further explained.

Errors are generally in two basic categories: systematic or random. *Systematic errors* conform to known mathematical and physical laws. They remain the same under set conditions. The sign is known but the magnitude varies with conditions. They generally change in proportion to the change in the condition causing the error. For example, a steel tape will stretch in proportion to the tension applied to it. Also, since the stretch behaves according to predetermined physical laws, the stretch can be calculated after measuring the tension. (See Figure 2.1) As another example, if the line of sight in a telescope of a surveyor's level was inclined, it would stay that way until it was changed by adjustment or other maladustment. Thus, it would **remain the same under set conditions.** Since the level could be tested to discover this error, the direction of inclination (sign of the error) would be known. The magnitude of the error would **vary with conditions**. That is, it would change as the sight distance changed, and in proportion to the sight distance as well as the inclination per unit sight distance.

Systematic errors are generally caused by (1) maladjustments of the instruments which can be traced to either misuse, normal wear, or manufacture of the equipment, or (2) natural phenomena. As such, they can usually be discovered through instrument tests which are designed to isolate the possible error and by a knowledge of the behavior of the instruments in the natural environment. It should be obvious that most systematic errors are detectible *and* removable. More will be discussed on these points later.

A special type of systematic error is a *constant error*. This is one that remains the same in both magnitude and sign through a given series of measurements. An index error (misalignment of the zero mark on scales) is an example of a constant error. Another constant error would be the error in a horizontal angle measurement caused by errors in centering the theodolite over a station. The error would remain the same no matter how many times the same angle was measured during the set-up.

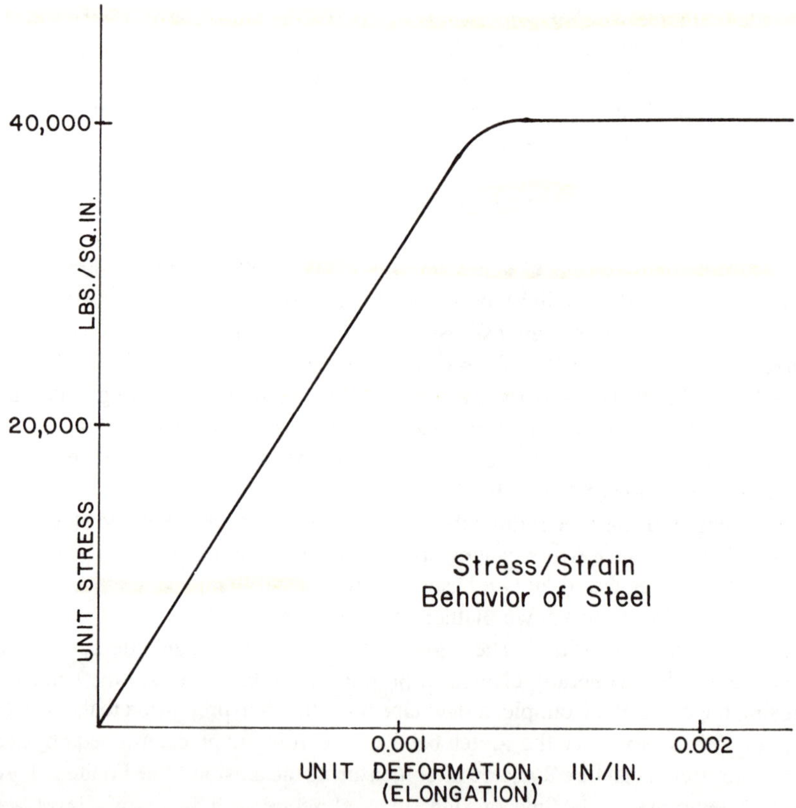

Figure 2.1 Effect of Tension on Elastic Material

Random errors are unavoidable. They can be minimized, but never eliminated. Such errors are sometimes called "accidental errors", but this term is less desirable since it is often confused with mistakes in meaning. Furthermore, "random" is more definitive of the nature in which the errors occur. The sign of a random error is never known. It is always plus or minus (written ±) in nature. The magnitude is unknown, and can be estimated only by experience or analysis. An example of a random error is the personal reading error of any scale. An observer estimates the final reading and could be either high or low in the estimation since exactness cannot occur. Random errors follow laws of probability and compensation. That is, large errors rarely occur, small errors are more likely than large ones, and there is an equal opportunity for positive errors to occur as negative ones. And, since the latter is true, they tend to cancel or compensate. Note, that they only *tend to* cancel. They never completely do so. More complete discussion of this tendency to cancel, as well as the estimation of the magnitude of random errors will be included in Chapter 4.

Understanding that random errors are unavoidable is a further important block in the foundation of building a complete professional approach to surveying measurements. The student must pause at this point to ponder inexactness and human nature. Mistakes are quite a different matter, but despite attempts to achieve perfection and even when mistakes are entirely avoided, random errors will always occur due to inability of people to achieve perfection in reading, manipulation, design, and manufacture of equipment as well as to unpredictable natural forces and instrument wear. If somehow, the perfect man-machine could be developed with powers sufficient to detect the slightest increment of error, random errors could be avoided. The presence of random or accidental errors has nothing to do with accidents or mistakes. Their magnitude may vary with care and training, but they will be present in the measurements of the most highly trained and conscientious individual who has a perfect record of no mistakes.

It should be obvious that the impossibility of avoiding random errors is at the root of the statement that **anything measured is inexact** (Section 1-2). The next section continues to place this in focus. The terms precision and random errors are closely related.

2-3 PRECISION AND ACCURACY

Just as the terms error and mistake are generally considered synonomous in layman terminology, precision and accuracy mean the same thing to most people outside of the world of measurement. Once again, the reader must make a conscious attempt to carefully absorb the definitions and explanations to follow. Comprehending the difference between the terms precision and accuracy is just as important as understanding the difference between errors and mistakes or between systematic and random errors. A sound understanding of these terms forms an important block in the foundation of one's development into a true professional in the world of measurement. In fact, faulty analyses of survey measurement data can result if the difference in these terms is misunderstood.

Precision means agreement among readings of the same quantity. If the readings are very close together in size, the set is more precise than if the values are widely scattered. Precision relates to the refinement (precision) in manufacture of the equipment and the care and refinement in making measurements. From this it can be deduced that precision and random errors have a very close relationship. Indeed, if the precision of the method is higher, the random errors should be smaller.

Accuracy means the agreement of the measurement or measurements (the average in a set) with the true value. A value that is closer to the true value is, by definition, more accurate than one which is farther from the true. Accuracy is a relative term when used in comparing measurements. Accuracy relates to how much attention has been paid to the detection and removal of systematic errors and mistakes. Accuracy and systematic errors go together in a manner similar to the way precision and random errors go together. It should be easily deduced that if readings were free from mistakes and all systematic errors were removed, such readings would be the absolute or true value (except for the effects of random errors). Less attention to removal of systematic errors and mistakes will result in less accuracy.

It is often said that precision relates to the method; accuracy relates to the result. The scatter of readings in a set tells something about the method, but it says nothing about the result as regards accuracy. A set of readings could agree very closely (be very precise), but be far from the true value because of a systematic error (actually, constant error) in the equipment and therefore be quite inaccurate. Likewise, another set of readings could have a lot of scatter and be less precise, but still be very accurate if the mean of the set or individual values closely agreed with the true value because of the absence of systematic errors.

An example will illustrate the difference between precision and accuracy. Suppose two surveyors each measured an angle six times and recorded their observations as follows:

Surveyor A		Surveyor B	
1	12° 14′ 32″	1	12° 14′ 30″
2	30″	2	35″
3	29″	3	28″
4	34″	4	30″
5	33″	5	36″
6	31″	6	32″

A brief study of the relative agreement among the readings in each set reveals that there is more *scatter* (dispersal about the average) in Surveyor B's data as well as a higher *range* (difference between largest and smallest values). The scatter and range are indicators of the precision since they reveal the relative agreement among readings in a set. Thus, it can be concluded that the set of measurements by Surveyor "A" has better *precision*.. Studying the definition of precision a little further, it could be reasoned that the care and refinement in making the measurements differed slightly and this caused the difference in precision. This difference in care and refinement probably didn't relate directly to differences in the type of instrument scale since both sets of data were recorded to the nearest second, but it could have related to skill of the observer, lighting conditions, optical features of the instruments or any number of other natural phenomena. Causes and sources of errors will be discussed subsequently—it is sufficient to recognize here that there was a slight difference in precision. Though the difference is perhaps not significant, observing it helps to understand the meaning of the term, precision.

From the definitions, it must be concluded that precision cannot be studied unless duplicate measurements (at least two) of a quantity are compared. But, since precision relates to method and equipment, intuition would dictate that more care in observing or a more finely divided scale would *probably* result in higher precision even without checking the agreement of duplicate measurements. This is not absolutely true, however, since there are factors involved other than observer care and scale divisions. Chapter 4 will contain explanations of how to check precision with more sophistication than was used in the above example.

Nothing can be stated regarding accuracy between the two sets of data unless the true value of the angle was known. This is an important point since accuracy is of more concern to the surveyor than precision. Results, that is, are what the surveyor must use in computations and show on maps. One could measure hundreds of repetitions of the angle and if a systematic or constant error was present the results would perhaps not be suitable (depending upon tolerances allowed) even with close agreement among the readings themselves. Therefore, surveyors should not misjudge that the measurements of Surveyor "A" are "better" than those of Surveyor "B" until something is known about either the true value or the extent to which each surveyor attempted to remove systematic errors. Suppose that the true value or a highly accurate value was known to be 12° 14′ 32.5″. Would not Surveyor B's set be "better" (more accurate) than A's? By definition, it would be. Surveyor A may have had a systematic error in centering the targets or instrument or a personal tendency to read a little low. Either instance would have caused a constant error in each value.

Just as a book's value can't be judged by its cover, likewise accuracy of results can't be judged by apparent quality in readings as reflected by precision. True quality is achieved only through thorough understanding of errors, their nature, possible magnitude, sources, and especially how to control them.

2-4 SOURCES OF ERRORS IN SURVEYING

Errors come from three sources: nature, instruments, and people. These are often categorized as natural errors, instrumental errors, and personal errors. It is helpful to understand these error sources when analyzing and dealing with errors.

Natural errors are caused by effects from nature, such as temperature, humidity, gravity, atmospheric pressure, atmospheric refraction, curvature of the earth, wind, tension, etc. Examples are the effect of temperature on a steel tape (it makes it expand or contract), or the effect of the atmosphere on a line of sight (it bends the light rays vertically downward). Natural errors can be either systematic or random. If the error follows a known physical law, it is systematic. If not, it is probably random. If an error cannot be isolated as either instrumental or personal, it is probably natural.

Instrumental errors are those caused by either (1) manufacture of the instruments, or (2) wear and/or maladjustments. The initial manufacture of the equipment affects the error magnitude and frequency of occurrence as it relates to the refinement in design and construction, quality control, etc. That is, some instruments are simply intended to be made better than others. They may have parts which are less susceptible to environmental effects, better precision in finishing of parts, smaller least count of scales, etc. So, when the instruments are new, the better ones (more costly, usually) would be expected to cause fewer or smaller errors than the ones made more crudely or cheaply. But, quality control enters into the picture and one could conceivably purchase a "lemon," even from a batch of precision instruments. This would be rare with most reputable manufacturers, however. Aside from manufacture, instruments do not always stay in adjustment and their parts may become worn or loose fitting. Even the best instruments are not completely immune to this. When instruments become worn or out of adjustment, errors result. Most errors resulting from manufacture or wear would be unpredictable and therefore, random in nature. The ones caused by maladjustment would generally be detectable and probably systematic. But, some errors due to manufacture would be systematic and some due to maladjustment would be random.

Personal errors are those deriving directly from people. Usually personal errors are caused by the inability of humans to perceive anything exactly. (Note: we are not talking about *mistakes* here!) This affects readings, aligning cross-hairs and other marks or centering devices. Aside from this innate inability which affects all of us, people vary in their manual dexterity, experience, training, intelligence, motivation, and desire to employ the appropriate care. Furthermore, fatigue, emotional fluctuations, and environment cause variations in personal errors. Thus, although nobody can measure exactly, some can undoubtedly do so better than others for several reasons. Personal errors are generally always random in nature, with one exception. The exception is that some people might have a tendency to read or align marks always to one side or another. This would cause systematic errors in results which may or may not be detectable. This result is called *personal equation* which could then be applied to measurements as a correction. Generally, however, this tendency, if it exists at all, is compensated through the measuring process. For example, if a timer at a race tends to have a slow reaction time at the gun and at the finish line, the error would cancel.

Students should attempt to categorize errors as above. The sources are sometimes confused. In categorizing, remember that it is the **source** that defines the error category. For example, temperature may cause an error in the instrument, but it is not an instrumental error since nature is the source. Reading a scale results in a personal error, but the inherent error due to the way the scale is made is an instrumental error.

2-5 DIRECT VS. INDIRECT MEASUREMENTS

If a measurement is made directly; that is, by measuring directly between points or lines, the measurement is said to be a *direct measurement*. If a distance, angle, or direction is computed from other measurements, it is said to be an *indirect measurement*. For example, a distance can be indirectly determined in a triangle by computing it from the law of cosines, employing measurements of the other two triangle legs and the included angle. A measurement made directly from a map or photograph becomes an indirect measurement of the ground distance when converted for scale. Distances measured electronically or tacheometrically (terms to be explained later) are sometimes said to be direct in that the distances are measured directly between two points, but are also called indirect in some texts since the distance is indirectly determined through various geometric or electronic principles built into the equipment. For the purposes of definition, computed quantities will be considered as indirect measurements. Values observed directly from the scales of instruments will be considered as direct measurements. As will be seen, nearly all measurements are indirect since computations (even of arithmetic means) are required to reduce them to a final single value of a quantity. This concept that nearly all measurements are indirect has much significance related to propagation of random errors, to be discussed in Chapter 4 and throughout much of the remainder of the book. The importance of the effects of computations (either in increasing or decreasing errors) on final quantities is often overlooked by practicing surveyors.

2-6 DEALING WITH SYSTEMATIC ERRORS

A correction is minus the error. That is, for every systematic error,

$$C = -E , \tag{2.2}$$

where C = the correction to remove the error, E. Applying this relationship to the definition of error,

$$T = R + C , \tag{2.3}$$

where

T = true value,

R = the reading.

In practice, there is probably more than one systematic error for which a correction must be applied, so the equation is more properly written as

$$T = R + C_T , \tag{2.4a}$$

where

C_T = Σ corrections or total correction.

This formula illustrates one way in which systematic errors can be counteracted. Another way is to compensate for them by measuring in such a way that the errors (and thus the corrections) are zero. When this is done, the formula reduces to T = R. That is, whatever is read is the correct or true value. An example is applying just the right tension to a tape so that its length is nominal (no correction needed). Another example is the adjustment of instruments. If the various axes are in their correct relationship, there is no instrumental systematic error from this source. It is not always practical to deal with natural systematic errors this way and a correction is usually necessary. For example, it would be impractical to wait until the outside temperature is that which makes the tape exactly its nominal length. Taping is done at whatever the temperature happens to be and a temperature correction is applied.

In theory, systematic errors can be counteracted or compensated. In practice they are never completely eliminated since the variable causing the error (temperature, adjustment, etc.) is itself subject to error in reading or aligning. The error remaining after counteracting or compensating is usually very small, but its size depends upon the precision of the method used to remove the error. This "residual" error is random in nature since it is assumed that the readings taken to remove the systematic errors involved only random errors.

Equations 2.2–2.4 illustrate a point about the definition of surveying. Recall that there are two fundamental problems in surveying: (1) determine relative positions, and (2) establish points. The first problem calls for solving the equation $T = R + C_T$. The second problem calls for solving the equation

$$R = T - C_T \tag{2.4b}$$

In the first problem the points, lines, or planes are fixed and readings are taken between them. In the second problem, the surveyor must solve for a reading, given the true value and the error information. The student should learn to grasp these elementary fundamentals. It will save much future mental strain since the algebraic signs work themselves out. A bad habit which many surveyors get into is thinking of effects of errors as "high or low" or "too small or too large," then attempting to manipulate readings using this rather unscientific logic. Invariably, mistakes are made in the algebraic signs of the corrections using this approach. Strict use of the formulas avoids this. This point is illustrated in several examples in Chapter 5.

The methods of dealing with random errors will be discussed in Chapter 4.

2-7 THE KEY TO UNDERSTANDING MEASUREMENT

All of the error related terms provide an avenue to the understanding of measurement—the world of the surveyor. The grasp of the idea that measurements are unlike discrete counts (exact) raises the professional to a level above the layman. Dealing with measurements in a logical way requires alteration of thought processes which have been formed as strong habits in many of us. For example, the popular phrase "to err is human," or "it is human to make mistakes" means the same in everyday language, but must take on different meaning to a surveyor. It is human to make both *errors* and *mistakes*, but the former must be minimized and the latter must be avoided in surveying. Surveyors cannot rationalize measurement mistakes on the basis of this quaint reference to human weakness and thereby use it as an excuse when mistakes are made. It *is* "human" to *err*, however, and the surveyor need not make excuses or apologies for that. Likewise, precision and accuracy mean different things to the surveyor. The student must alter the habit of thinking of them as synonymous. Much of surveying (or any) education is the molding of thought processes while discarding misconceptions which have formed into habit. Learning survey measurement theory follows this process.

Problems - Chapter 2

2.1 Two people paced a distance two times each and from a count of their paces and estimate of length per pace, pacer "A" observed values of 216 and 219 feet and pacer "B" observed 215 feet and 221 feet. Which pacer had better precision?

2.2 If, in Problem 2.1, the distance was later taped and found to be 218.2 feet, whose data were more accurate, pacer "A" or pacer "B"?

2.3 Distinguish between random error and mistake.

2.4 Classify the following as to error source:
a) Sag of a surveyors tape when held horizontal and unsupported by the ground.
b) Estimation of ½ degree when measuring with a protractor.
c) Error due to uneven graduations on a protractor.

2.5 Explain how accuracy increases when systematic errors are reduced.

2.6 Explain how precision increases when random errors are minimized.

2.7 Why would a finer divided scale system probably yield more precise measurements?

2.8 Name a situation in which the true value is known in surveying.

2.9 Differentiate between direct and indirect measurements.

2.10 While dressed, a person read his weight from a scale as 140 lbs. His clothes and shoes weighed 2.3 lbs. and later it was discovered that the scale read 148.5 lbs. when three calibrated 50 lb. blocks were placed on it. What is the man's true weight? (Solve by first identifying and evaluating the errors and corrections.)

Additional Reading References — Chapter 2

1. Barry, B. Austin. *Errors in Practical Measurement in Science, Engineering, and Technology*. New York: John Wiley & Sons, Inc., 1978, Chapter 2.

2. Mikhail, Edward M. and Gracie, Gordon. *Analysis and Adjustment of Survey Measurements*. New York: Van Nostrand Reinhold Company, 1980, Sections. 1.1-1.2.

CHAPTER 3

SURVEYING INSTRUMENTS AND THEIR USE

3-1 INTRODUCTION

The intention of this chapter is to develop an understanding of basic surveying instrument construction and application and the mechanics of surveying measurement so that the student can relate to the material in the remainder of the book. Detailed facets of field surveying and instrument manipulation will be left as subject matter for another book. Instruments to be described are typical of those used by modern surveyors in their everyday work. Exotic, specialized, or obsolete equipment is omitted, the inclusion of which would detract from the central purpose of this book.

3-2 COMMON COMPONENTS OF SURVEYING INSTRUMENTS

Telescopes

The telescope was invented in 1608. It was first used in surveying instruments in 1669. Its use in surveying did not become widespread until the late eighteenth century. Since then, it has gradually replaced the simple peepsight as a means of defining a straight line in surveying.

The purpose of any telescope is to magnify or enlarge the image of an object. In surveying, the purpose is also to define a line of sight. In older instruments, before telescopes were used, the line of sight was defined by peep-sights similar to what are used on rifles. Any hunter will recognize the advantages of a telescopic sight on a rifle. The advantages in surveying are the same.

The main parts of a telescope are the objective lens, eyepiece, and reticle. The *objective lens* is the large lens at the front of the telescope. It is generally a compound lens, being either a double or triple lens system. Light rays striking this lens system form an image within the telescope. Generally, a larger diameter objective lens receives more light, resulting in brighter images.

The *eyepiece* is the small lens at the viewer's end of the telescope. It is a lens system consisting of either two or four lenses. The four-lens system provides for erecting the image which would otherwise be inverted. The eyepiece is a microscope which permits the viewer to see the image formed by the objective lens and which magnifies the image.

The *reticle* (sometimes spelled reticule) is often called the "cross-hairs," a more descriptive term for this system of lines which define the line of sight and permit precise pointing and reading through the telescope. The reticle is mounted near the eyepiece end of the telescope, at the principal focus point of the objective lens. Physically, the reticle is a system of wires, hairs, threads, or lines etched on glass. In modern instruments, it is generally the latter. Cross-hairs were first placed in telescopes in 1640 and were then threads of silk. The reticle contains at least a horizontal and a vertical line. A straight line connecting the point of intersection of the cross-hairs with the optical center of the objective lens define the *line of sight* (sometimes called line of collimation) of the telescope. In addition to the horizontal and vertical cross-hairs, the reticle may contain additional lines for special measurements. Some of these will be explained later. Reticles are illustrated in Fig. 3.24.

Modern telescopes also contain a *negative lens*. This lens is mounted between the reticle and objective lens in such a way that it slides between them when the *focussing knob* is adjusted. The purpose of the negative lens is to focus the objective onto the reticle for clear viewing. When such a lens is employed, the telescope is said to be *internal focussing*. The objective lens does not move during focussing. Such a telescope is completely enclosed and this gives greater protection against dust and moisture. Older surveying instruments sometimes will still be found which have an *external focussing* telescope (sometimes called astronomical telescope). This type has no negative lens. Instead, the objective lens itself moves when the focussing knob is adjusted. Such telescopes are longer than the internal focussing type. Since there must be an external focussing sleeve, this type of telescope is more susceptible to moisture and dust penetration into the system. Views of both types are illustrated in Figs. 3.1 and 3.2.

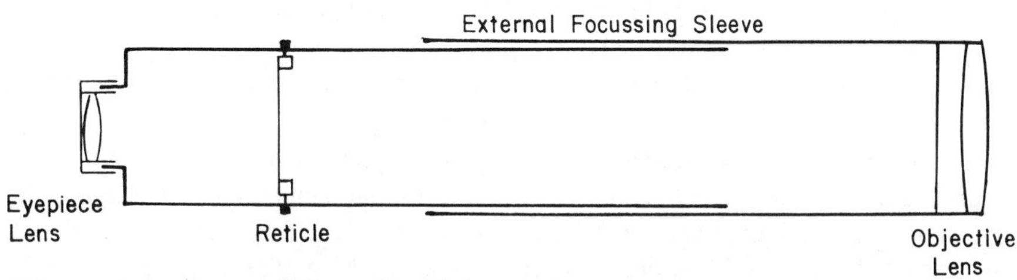

Figure 3.1 External Focussing Telescope

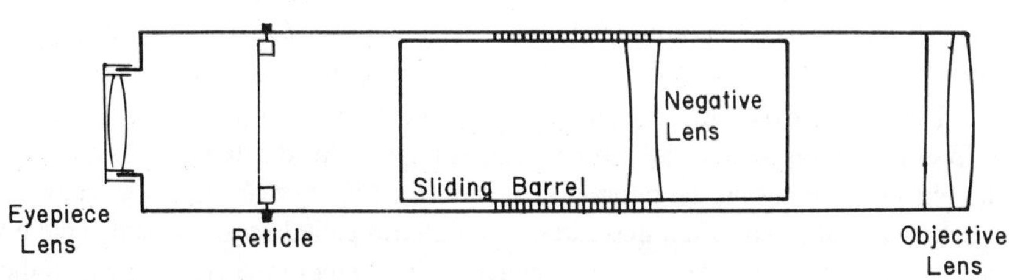

Figure 3.2 Internal Focussing Telescope

Most telescopes used in surveying have a *magnification* between about 20 and 40 power. This is a range which provides ample magnification for distances and target sizes normally encountered in surveying and in which the *definition* or clarity of the objective is optimum. Magnification is the ratio of the apparent size of an object as seen through the telescope to its size as seen by the unaided eye. Definition is determined by factors such as the quality of the optics and the amount of light received through the objective lens and brought to the viewer's eye. Definition is the quality that directly affects *pointing accuracy*, which is the exactness with which the line of sight can be pointed to a target. This will be discussed further in Chapter 6.

To focus a surveying telescope, the eyepiece is first focussed on the reticle plane which results in a clear image of the cross-hairs. This adjustment should remain fixed for an individual, but will vary with each person because of differences in eyesight. Objects (rods, targets, etc.) are focussed as sharply as possible using the objective lens focus knob. Focus of objects changes with object distances. When the telescope is properly focussed, both the cross-hairs and object should be clear and there should be no apparent movement between them if the observer shifts his head slightly. If such apparent movement occurs, *parallax* exists, which can only be corrected by proper focussing. Parallax will be discussed further in Chapter 6.

Telescopes with various types and styles of reticles are important components of surveyors' theodolites and levels. An excellent discussion of the optics related to surveying telescopes can be found in Reference 2.

Level Vials

Water at rest in troughs, glasses, tubes, and other receptacles has been used since ancient times to determine horizontal directions. With gravity alone acting on the water, its surface is horizontal. The earliest version of a *level vial* (also called bubble tube) was introduced in 1666. It was a simple enclosed tube filled with water. Later developments resulted in internal curvature of the upper tube surface and the water was replaced with ether or some other fluid with a freezing point lower than water. Modern vials contain purified synthetic alcohol which is quick acting and relatively stable in length for normal temperature variations. All vials have a small air space (the bubble itself). The curved top surface provides a means of stabilizing the bubble; otherwise it would be quite erratic. The curved upper surface is a segment of a circle which has evenly marked divisions etched on it. The markings are generally 2 mm long but some instruments have 0.1 inch or 0.01 foot divisions. Centering the bubble is accomplished by aligning its ends at equal distances from the midpoint of the tube using the marks as guides.

The *bubble axis* is the line tangent to the upper inside surface of the tube at its midpoint. The midpoint is the geometrical center between the divisions as marked. The object of centering the bubble is to cause the bubble axis to be horizontal.

Bubble tubes are often described using a term called *sensitivity* (also called sensitiveness) which is the angle subtended at the center of the circle of which the upper tube surface is a segment, by a change in position of one division of the bubble. Sensitivity is expressed in seconds of arc. It can be seen that sensitivity is a function of the radius of curvature of the tube as well as the size of the divisions etched on the tube. It should be obvious that a longer radius of curvature would result in a higher sensitivity. But, the actual numerical value, in seconds, would be smaller. Bubble tubes on typical surveying instruments vary in sensitivity between about 10 and 30 seconds. Generally speaking, the higher the sensitivity (10'' is 'higher' than 30''), the more precise the instrument. But sensitivity which is too high can be a nuisance for rough survey work because of the added time and difficulty required to keep it centered. The geometry of the bubble tube is illustrated in Fig. 3.3. In this figure,

$$\tan \mu = \frac{b}{R_b} = \frac{j}{D} \tag{3.1}$$

where

μ = bubble sensitivity, seconds

b = value of one division of the bubble

R_b = radius of curvature of the inside, upper surface of the bubble

j = deviation of the bubble axis direction at distance, D

D = distance from bubble center to any point or target.

It should be obvious that $\tan \mu = \sin \mu = \mu$, radians, since μ is very small.

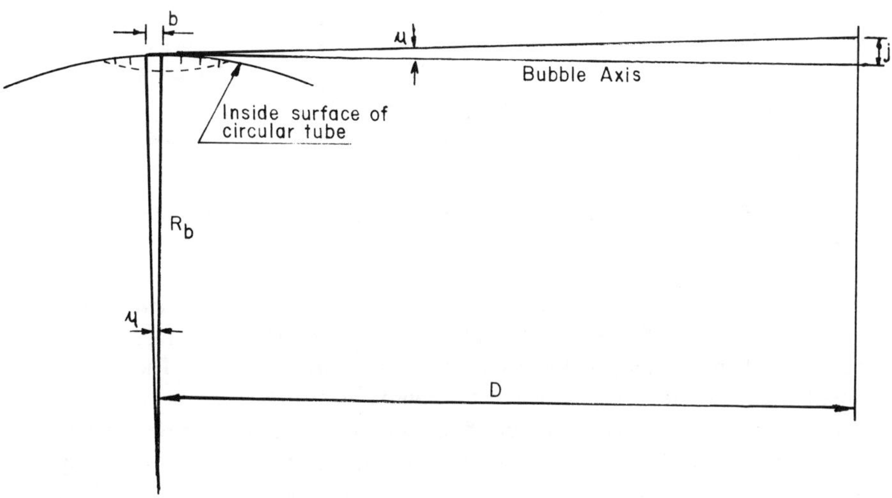

Figure 3.3 Bubble Sensitivity

Some surveying instruments employ a *coincident bubble*. The operator is provided with a means of viewing a split image of the two ends of the bubble. This is accomplished by a prism which makes the two ends visible simultaneously through an optical microscope. The instrument is constructed in such a way that the bubble is "centered" when the two ends are adjusted to align as a smooth arc as shown in the center of Fig. 3.4. An adjustment knob is provided on the instrument for this aligning. The image is magnified so that centering is accomplished more precisely than when using a non-coincident type bubble. Manufacturers state that the coincident bubble can be centered about four times as precisely as a non-coincident type bubble. As will be seen when elevation instruments are discussed, another advantage of the coincident bubble system is that the optical microscope used to view the bubble image is mounted where the operator can view the bubble almost simultaneously with viewing an objective through the telescope.

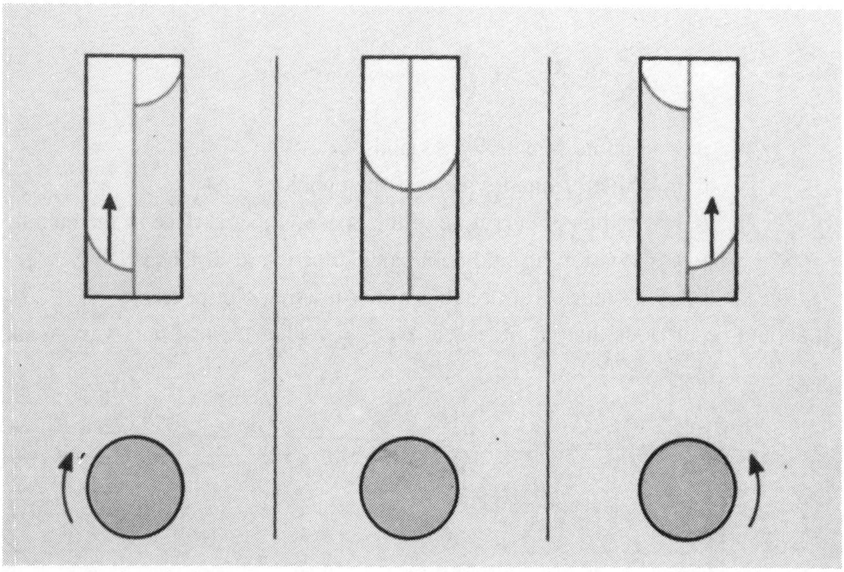

Figure 3.4 Coincident Bubble (Courtesy of Wild Heerbrugg Instruments, Inc. Farmingdale, N.Y.)

Another type of level vial used on survey instruments is a *circular bubble* (sometimes called ''fish-eye'' or ''bull's eye'' bubble). The top surface of such a vial is a sphere and the bubble formed by the air space is a circle. The vial has a small circle etched on its surface of the approximate size of the air bubble. The ''axis'' of such a vial is a plane tangent at its center. When the bubble is centered within the etched circle, the axis plane is horizontal. Such vials are not generally as sensitive as longitudinal bubbles. They are used for purposes of rough leveling and plumbing rods or poles. Their advantage is that horizontal alignment is established in all directions instead of just one direction.

Common surveying instruments have a vertical axis (axis of rotation). If an instrument is constructed and adjusted in such a way that the bubble axis is set perpendicular to this rotation axis, then the latter axis will be truly vertical when the bubble is centered in all positions around the axis of rotation. Another example of the use of the level vial is a surveyor's level. If such an instrument is constructed and adjusted so that the bubble axis is set parallel with the line of sight of the telescope, then the line of sight will be horizontal when the bubble is centered.

As should be obvious, the level vial, along with other types of bubble instruments, is another important component of surveying instruments. Also, the bubble tube and the circular bubble are used as separate instruments for a variety of problems to establish horizontal and vertical alignment.

Plummets

A *plumb bob* is a weight, usually conical in shape and made of brass or iron, suspended by a cord, by means of which a point can be projected vertically over relatively short distances (see Fig. 3.5). Plumb bobs vary in weight from about 6 to 16 ounces (170-450 grams). They have been used since ancient times. One application is for defining a vertical line, such as in plumbing a building column or vertical mining shaft. A more basic use is in centering a survey instrument over a ground point. The plumb line (the cord attached to the bob) is often used as a target for sighting with angle measuring instruments.

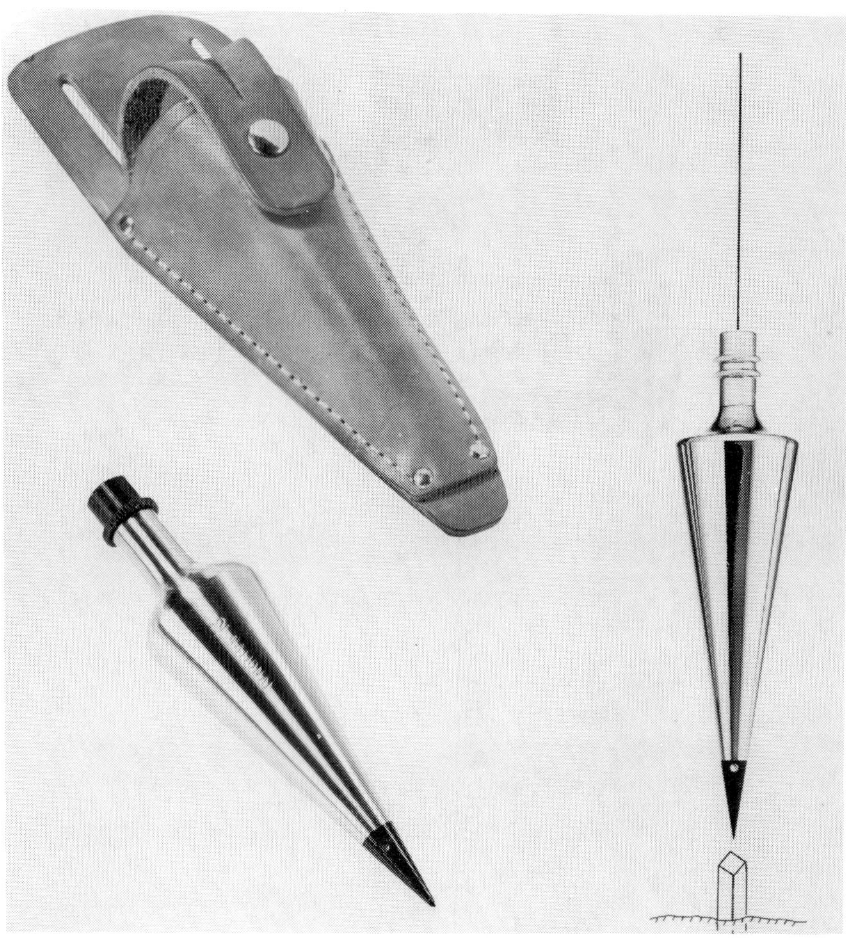

Figure 3.5 Plumb bob (Courtesy of Chicago Steel Tape Co., Chicago Heights, Ill.)

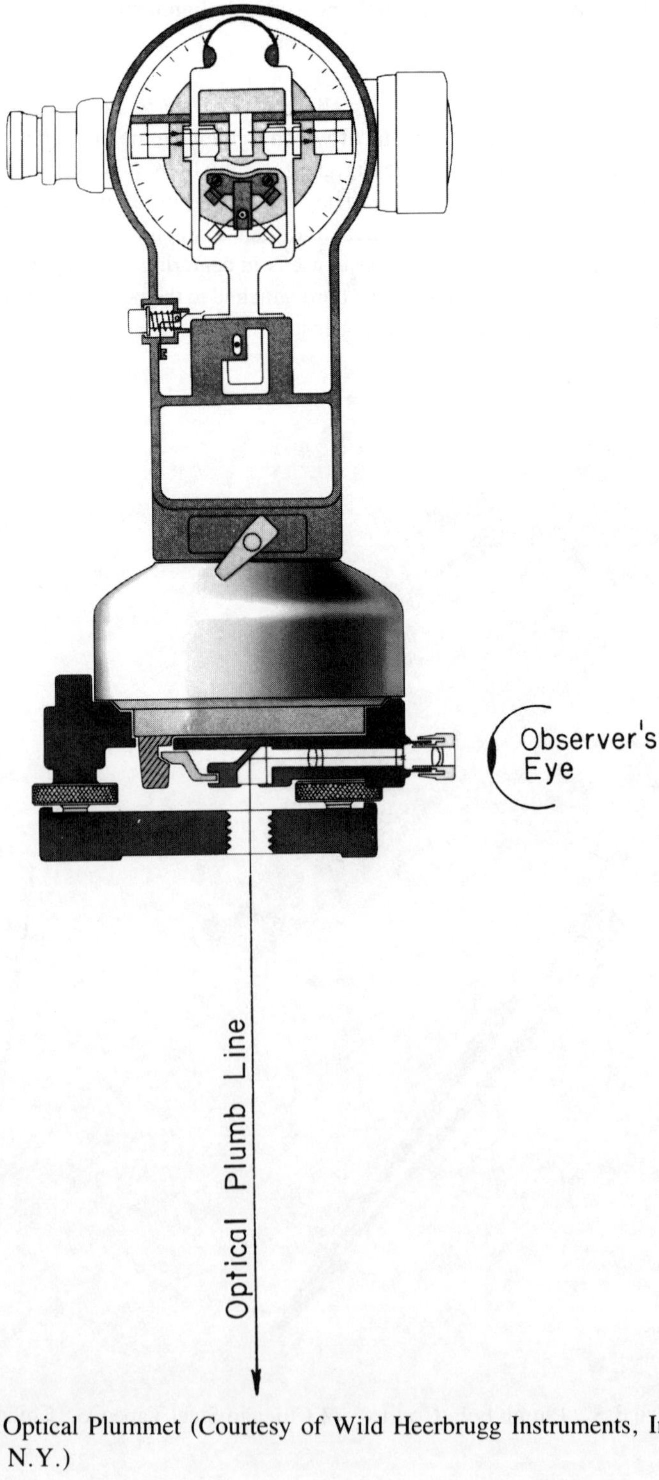

Figure 3.6 Optical Plummet (Courtesy of Wild Heerbrugg Instruments, Inc. Farmingdale, N.Y.)

Modern survey instruments usually have an *optical plummet,* a device for optically centering over a ground point without the need for a plumb bob. This is a simple 90° prism mounted in the base of the instrument in line with the instrument's vertical axis. An eyepiece and reticle provide the means to view the ground point. (See Fig. 3.6) After leveling the instrument, the observer focuses the eyepiece and looks horizontally through it, viewing the ground mark along the vertical line passing through the instrument's vertical axis. Optical plummets have optical characteristics suitable for centering over points at normal tripod height (1-2 meters). The obvious advantages of optical plummets over plumb bobs are that they are unaffected by wind and require much less hand manipulation.

A *vertical collimator* is a telescope mounted such that its collimation axis can be made to coincide with the vertical. This instrument is similar to an optical plummet except that it has telescopic power suitable for plumbing an instrument over a target from a high tower.

Another plumbing device used on some modern instruments consists of a rod attached at the base of the instrument so that its axis aligns with the instrument's vertical axis. It has a tip which is placed on the ground point. It "telescopes" so that it can be adjusted for slight differences in tripod height. A circular bubble is affixed to it, and when this bubble is centered, the rod is vertical and thus, the instrument is considered centered over the point. Furthermore, the tripod head is nearly level when the rod is plumbed, making final instrument leveling quicker. This plumbing device is shown in Fig. 3.7.

Scales and Verniers

Surveyors and surveying technicians must become skilled in reading various scales. Students are probably already familiar with reading the several types of engineering scales used for drafting. Most scales found on surveying instruments are principally the same, being subdivided into units, tenths, and possibly hundredths or thousandths. Readings are generally made by observing the nearest mark or estimating to the tenth between marks. The art of reading scales consists of careful alignment of index marks, assurance that proper illumination and magnification is used, and quick but careful reading of the finest divisions of the scale. Most surveying scales are either linear or circular. Illustrations of several types of scales are included with the following discussion of verniers and with the descriptions of survey instruments.

A *vernier* is an auxiliary scale sliding against and used in reading a primary scale. The vernier was invented by Pierre Vernier in 1631. Its purpose is to enable reading a principal scale (such as a divided circle) much closer than one division of that scale. If a division on the primary scale is longer than a single division on the vernier, it is a direct vernier. If a division on the vernier is longer, it is a retrograde vernier, so called because its numbers run in the opposite direction from those on the primary scale. The direct vernier is the usual type. Two verniers extending and numbered in opposite directions from the same index form a double vernier, used in reading a circle having graduations numbered in both directions. A single vernier so constructed and numbered that it may be read in either direction is termed a folding vernier.

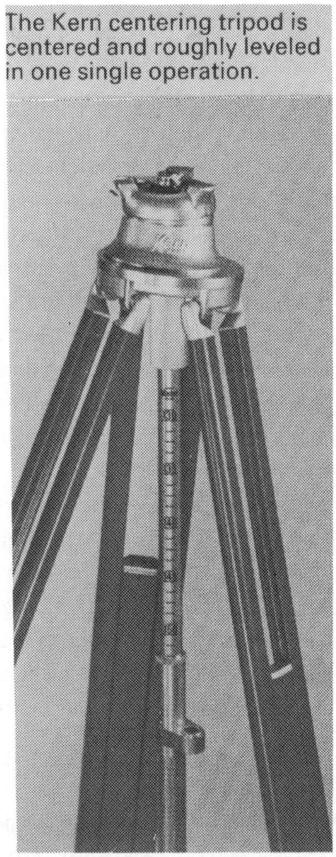

Figure 3.7 Kern Rod Plummet (Courtesy of Kern Instruments, Inc., Brewster, New York.)

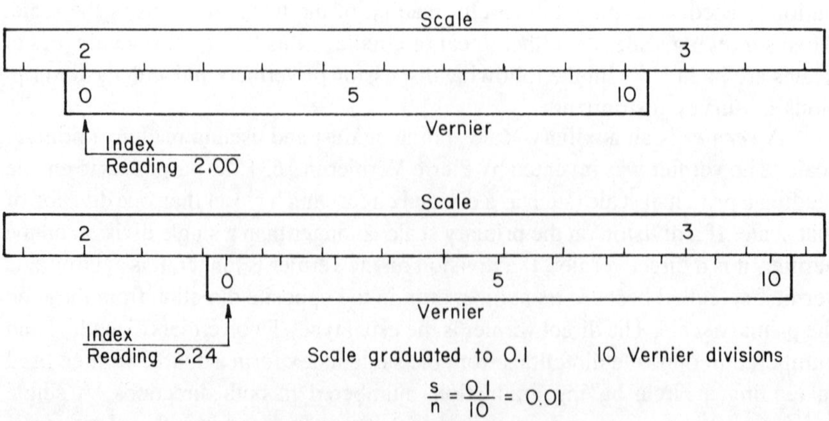

Figure 3.8 Linear Scale and Vernier

Figure 3.8 shows a linear scale with vernier. The vernier is divided into 10 parts covering 9 parts on the scale. The value of one vernier division is equal to 0.9 of one scale division. In this figure, the *index* (zero mark) of the vernier strikes the scale between 2.2 and 2.3. Even without the vernier, it appears that the reading is around 2.23 or 2.24. This last digit (hundredth) is more precisely determined from the vernier where the marks align with the scale divisions. The alignment occurs most closely at 4 on the vernier; therefore the reading is 2.24.

The smallest division that can be read from this scale-vernier system is 0.1 of the scale division. The general principle of any vernier is

$$D_s = \frac{s}{n} \tag{3.2}$$

where

D_s = smallest scale division that can be read
s = the value of a scale division
n = the number of divisions on the vernier.

A typical circular scale is divided into half-degrees or 30 minutes as shown in Fig. 3.9. The accompanying vernier has 30 divisions. Thus, from the above equation, this would be a 1 minute direct reading vernier. It will be noted that this is a double vernier, which permits readings to be made either clockwise or counterclockwise.

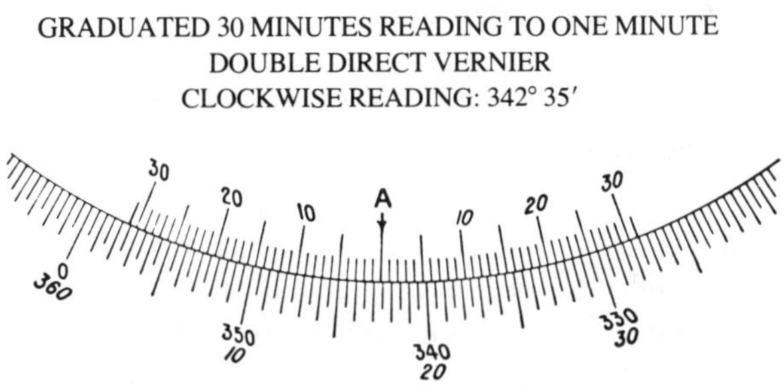

Figure 3.9 Circular Scale and Vernier (Used with permission of Keuffel & Esser Company, Morristown, New Jersey.)

A *micrometer* is an instrument, used with a telescope or microscope, for measuring minute distances. The measurement given directly is that of the image of the object formed at the focus of the object glass. Micrometers are used on modern angle measuring and some other surveying instruments. In angle measuring instruments, magnification of the micrometer is 40 to 70 times and scale divisions are etched precisely on circular drums. Such scales can be manufactured and read very precisely.

The term *least count* means the finest reading that can be made directly (without estimation) from any scale, scale-vernier system, or micrometer. It is very important that any surveying technician first determine the least count of a measuring system before attempting to make readings.

Tripods and Staffs

A tripod is used with most principal surveying instruments for support and maintaining stability during observations. Surveying tripods have an upper element (head) which is a smooth plane surface for mounting the instrument, a device such as a threaded bolt for securely attaching the instrument to the head, and a means of making lateral adjustments of the instrument on the head of the tripod. The legs are either metal or wood and are hinged at the head. The legs have pointed metal shoes which can be pressed into the ground by pushing on spurs attached to the shoes. Tripods are either fixed leg (non-adjustable in length) or extension leg (adjustable). The tripod should be considered as an important part of a measuring system.

Simple vertical staffs can sometimes be used for supporting light instruments. Naturally, they do not afford the lateral stability of a tripod. Staffs are used only for rough measurements. They have been used for supporting compasses and other instruments since medieval times.

Rigid supports are used in highly precise work and when adjusting instruments. Such supports are more permanently constructed than tripods and consist of either concrete columns supported by additional concrete platforms or a truss type frame with a mounting head which is similarly mounted atop several feet of concrete. Such a rigid support is shown in Fig. 3.10 where an instrument is undergoing adjustment in a laboratory.

3-3 ANGLE, DIRECTION, AND ALIGNMENT INSTRUMENTS

Angle and Direction Measurements

The various common angle measuring instruments for surveying are, in concept, similar to the common protractor. If the index (0° mark) of a protractor is aligned along a straight line and centered at a point on the line, an angle can be observed to another line from the same point. Also, another line can be established with respect to the direction of either of these existing lines. Field work is done over areas much larger than a piece of plotting paper. Instead of constructing giant protractors for field use, early surveyors and mathematical practitioners developed instruments which have a circular scale (a form of protractor) and a method, such as a peepsight or telescope with a reticle to precisely extend a line past the actual physical mark on the circular scale. Such instruments are discussed in the next section.

Historical Development

The *theodolite* is the most commonly used angle measuring surveying instrument. The word probably originated from "theodilitus", which was a term used to describe the measuring circle of a sixteenth century instrument. Theodolites used through the sixteenth and most of the seventeenth centuries employed open sights. A practical telescopic instrument was not developed until 1720. Theodolites used through most of the nineteenth century looked much like the mid eighteenth

Figure 3.10 Laboratory Adjustments of a Theodolite

century instrument shown in Fig. 3.11. These were called *altazimuth theodolites*, the term deriving from the fact that both altitude (vertical angle) and azimuth (horizontal angle) could be measured. Exceptions were *geodetic theodolites* used for precise work which were much bulkier.

During the 1830's a variation of the altazimuth theodolite began to appear. The standards (supports for the telescope) were raised, the telescope was shortened, and thus it could be rotated about an axis perpendicular to the line of sight. This

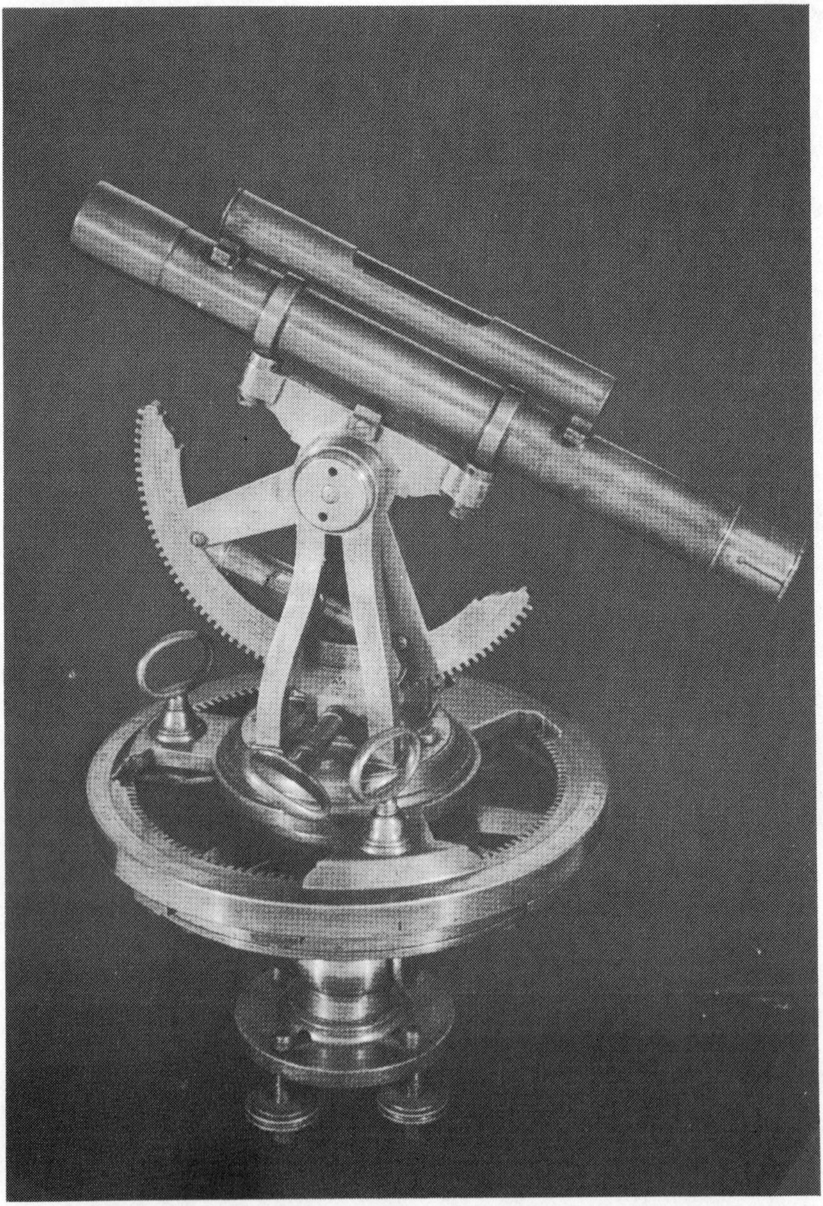

Figure 3.11 Eighteenth Century Theodolite (Courtesy of Crown Copyright, Science Museum, London.)

form of theodolite became known as a *transit theodolite* since the telescope could be "transited" through the zenith (inverted and pointed in the opposite direction). The circular scales were etched on metal plates and read with the aid of verniers.

In the early twentieth century, Heinrich Wild, of Switzerland, began to develop some of the first improvements in theodolite reading scales. Gradually, a transit theodolite which could be read with the aid of an optical micrometer was developed. Circles began to be made of glass rather than metal to guarantee more uniformity of the circle graduations and better reading quality. Besides improving reading methods and precision, Wild also decreased the size and weight of the theodolite and improved its resistance to weather and shock. Because of added precision in the glass scales and optical reading method, the vernier is not needed on such instruments. Furthermore, circles are much smaller and enclosed and thus not visible directly as with the older vernier type instruments. Because the circles of these theodolites are read with optical microscopes, they are called *optical transit theodolites*.

Research and development of optical theodolites has continued mostly in Europe and more recently in Japan. American manufacturers have continued to produce the more cumbersome and less precise vernier transit theodolites. Wild's developments had remarkably little effect on most surveyors in the United States and even now, many present practitioners still favor the vernier transit theodolite. There is a trend toward use of the optical type, however, a trend which has been slowed somewhat because of radical increases in prices of the optical type during the 1970's.

The essential features of any theodolite are shown in Fig. 3.12. Note that each instrument has both a vertical and a horizontal circle serving in a manner similar to a protractor, the line of sight is defined by a telescope reticle and objective lens, and axes of rotation provide for measurements completely around a circle. More details in function and operation will be explained subsequently.

Confusion in Terminology

Confusion exists, particularly among American surveyors, concerning terms used to refer to angle measuring instruments. As has been mentioned, the term "altazimuth theodolite" was used to describe the eighteenth century instruments, followed by "transit theodolite" after the transiting feature was added. Sometime after this, American surveyors apparently abbreviated the term to simply "transit." Then, when the optical theodolites were developed decades later, they became known simply as "theodolites" in American surveyors' jargon. Actually both the vernier and the optical type are "transit theodolites." Proper terminology would be "vernier transit theodolite" and "optical transit theodolite," the main point being that both types are theodolites, the word "transit" being used to distinguish them from the much earlier instruments whose telescope could not be plunged and the word "vernier" or "optical" being used to describe the reading method. The latter type is made in a variety of ways resulting in additional descriptive terms such as double centering theodolite, scale reading theodolite, coincidence type theodolite, and such other manufacturers' terminology. Attempts will be made to explain these differences through the classifications and illustrations to follow. Practicing surveyors sometimes classify the less precise optical theodolites as "optical transits." This has probably been caused by associating the repeating feature (to be explained in the next paragraph) with the American made

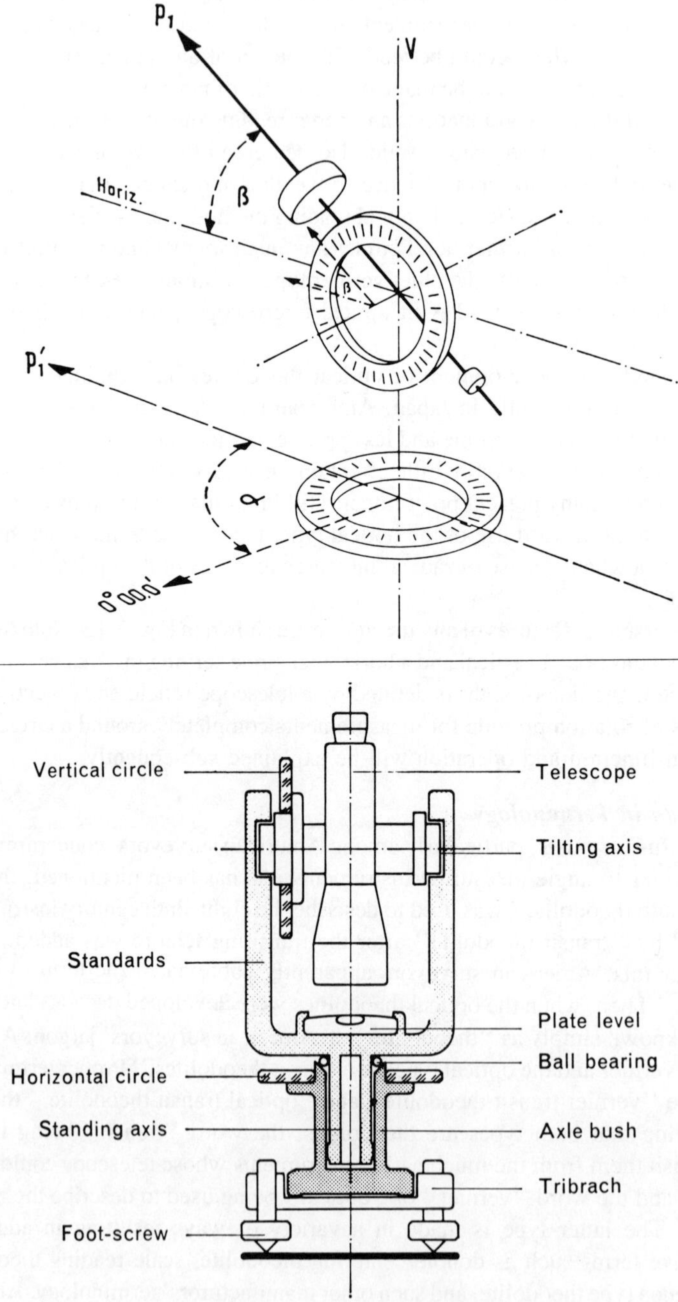

Figure 3.12 Principal Features of a Theodolite (Courtesy of Wild Heerbrugg Instruments, Inc., Farmingdale, N.Y.)

Figure 3.14 Vernier Theodolites (Transits)

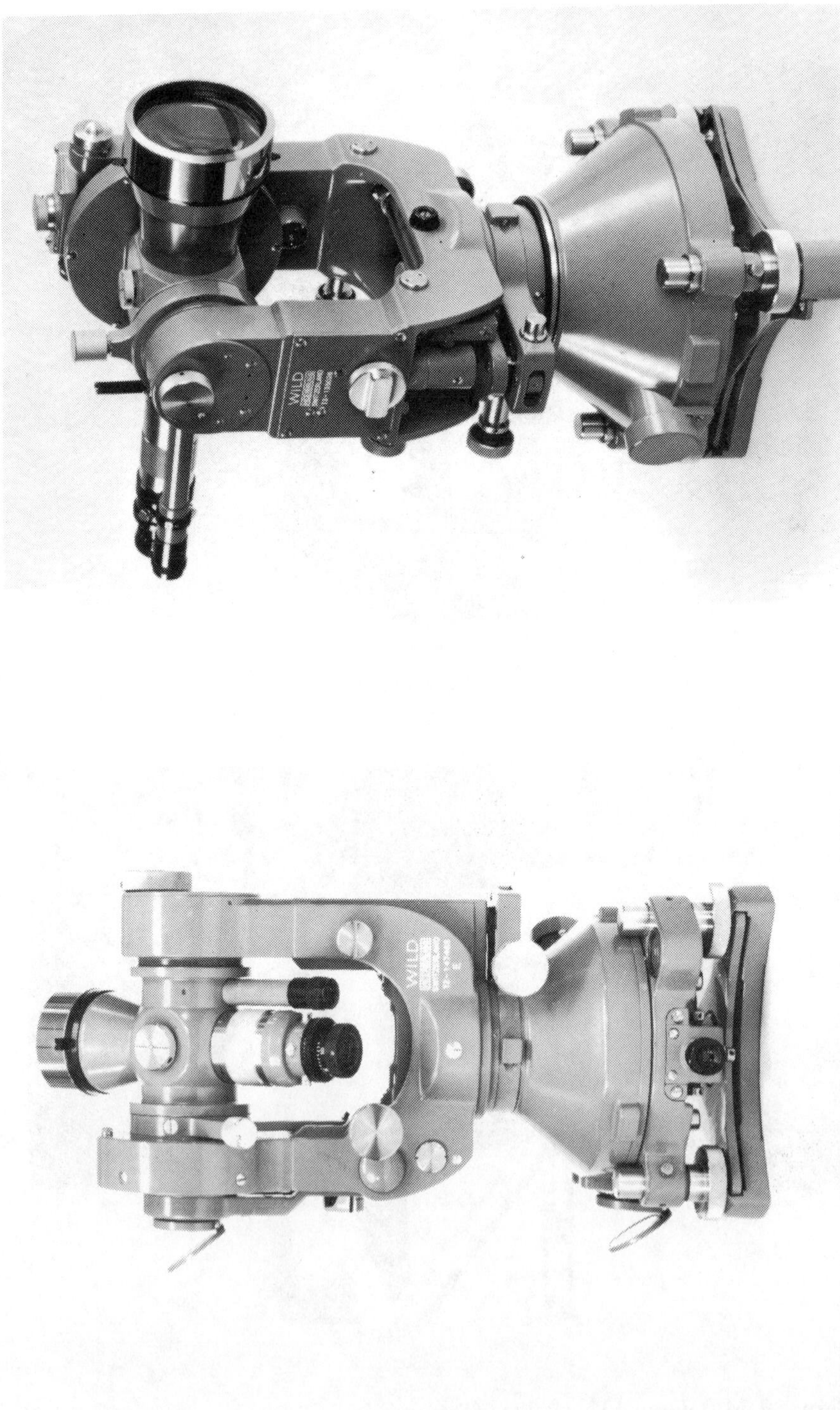

Figure 3.13 Direction Theodolites (Courtesy of Wild Heerbrugg Instruments Inc., Farmingdale, N.Y.)

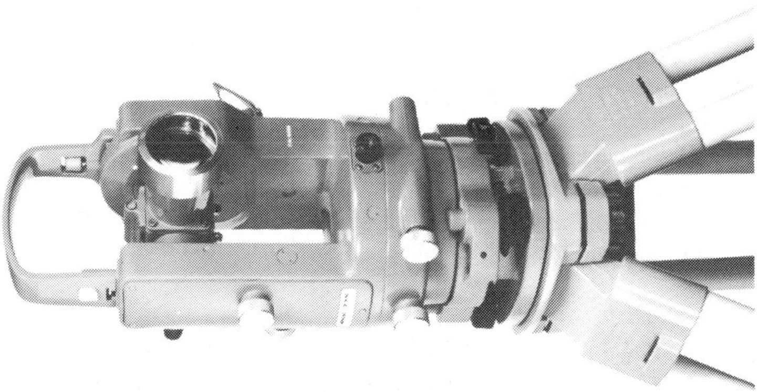

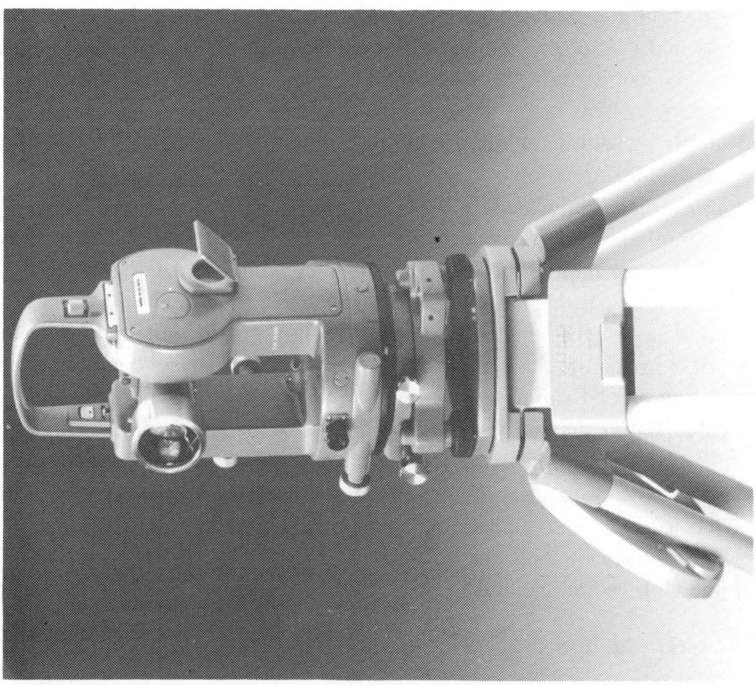

Figure 3.15 Optical Repeating Theodolites (Courtesy of Wild Heerbrugg Instruments Inc., Farmingdale, N.Y.)

"transits" and since these lower precision optical instruments operate similarly, the logic is that they must also be transits. This same group of surveyors would then identify the very precise geodetic theodolites (direction type—to be explained) simply as "theodolite."

It is to be understood that all such modern instruments, whether read with a vernier scale or through an optical miscroscope with a coincidence circle, are theodolites, technically speaking. Furthermore, they all "transit" so that distinction is unnecessary nowadays. It would seem proper to refer to them as either vernier theodolites or optical theodolites, further breaking the latter into categories as needed according to differences in reading systems and other features.

Other jargon occasionally heard in some countries to distinguish the optical from the vernier type theodolites is "glass circle" (optical) or "metal circle" (vernier). These terms are inappropriate since some Japanese models use glass verniers with glass circles.

Classification of Instruments

The type of instruments which have evolved through the efforts of Wild and others are for two broad classes of surveying, one being precise geodetic surveying and the other being what might be termed "ordinary" surveying. This separation into classes is not distinct, however, since geodetic instruments may be occasionally used for a combination of surveying operations on a particular project and some of the less precise theodolites are used for lower precision geodetic surveys.

The more precise geodetic instruments fall into a class called *direction theodolites*. Instruments in this group have but one horizontal motion, with its clamp and fine motion screw. The horizontal circle remains fixed during a set of observations and cannot be clamped to the alidade (upper, rotating part) of the instrument. The name implies that **directions** are "observed" rather than **angles** being "measured." This helps to explain the principle. If an angle is desired, it is obtained simply by subtracting the direction readings to two targets. Examples of direction theodolites are shown in Fig. 3.13.

Repeating theodolites usually have an upper and lower horizontal motion clamp, each motion having a fine motion screw. This feature provides for rotating the alidade and circle about the vertical axis together, without changing the reading and for zeroing the instrument for angle measurements. In brief, the upper motion provides for measuring angles and the lower motion provides for repeating the angles multiple times for added precision. American made vernier theodolites are of the repeating type. Examples of vernier theodolites are shown in Fig. 3.14.

Optical repeating theodolites are manufactured with more than one design for clamping the horizontal circle to the alidade. The Wild T1, for example, has upper and lower clamps and fine motion screws similar to the vernier theodolites. The Wild T16 has a circle locking lever to clamp or unclamp the circle. The same fine motion screw is used for both the upper and lower motion on the T16. Optical repeating theodolites are shown in Fig. 3.15.

Functions of Theodolites

Horizontal angle measurement. The principal function of a theodolite is horizontal angle measurement. In reference to this function and other functions, certain terms should first be explained. There are two positions of the telescope, being either *direct* (also called erect) or *reverse* (also called inverted). On vernier theodolites the direct position is usually when the objective focussing knob is on top and the telescope bubble is beneath the telescope. On optical theodolites, the direct position is usually when the rough peepsight is on top of the telescope and the optical microscope is on the right as viewed from the observer's end. In most countries other than the U.S., the term *face right* and *face left* is used to explain the difference between the telescope positions. The term refers to the location (right or left) of the vertical circle as viewed from the observer's end. The telescope is reversed in position by rotating it 180° about the horizontal axis (transiting).

The instrument is first mounted on a tripod. Then with the aid of bubble tubes and either a plumb bob or optical plummet, the horizontal circle is placed in the horizontal plane and the vertical axis made vertical while passing through the marked ground station. Surveyors usually refer to this as "setting up" the instrument. When leveled and centered over the ground station, the theodolite is ready to measure angles as when a protractor is centered on a point marked on plotting paper. Next, the eyepiece of the telescope is focussed for the observer's eye.

Assuming the instrument is a repeating theodolite it is next set to zero by proper manipulation of the upper and lower motion. Now, it is ready to point to a target using the lower motion. After rough pointing, the lower motion is clamped, the objective lens focussed, and the vertical cross-hair precisely pointed on the target using the lower fine motion screw. Next, the upper motion is released and the telescope aligned to another target to which the horizontal angle is desired, using the upper fine motion screw for precise pointing. The angle is next read and recorded. To compensate for certain instrumental errors (to be explained in Chapter 6) as well as reduce effects of some personal errors, the angle should be repeated with the telescope in the inverted position. The final reading should then be twice the single reading plus or minus the effect of errors. The value used would be one-half the double reading. This value should theoretically be free from systematic instrumental errors whose signs are reversed when the telescope is inverted. The procedure just described is commonly called the *double angle procedure*. For added precision, angles are often repeated several more times being sure to have an equal number with the telescope direct as inverted. This is commonly called *angles by repetition*, the double angle procedure merely being a special case of this procedure where only two repetitions were made.

The angle measuring procedure is similar with a direction theodolite, the major differences being that the scale should not be set to zero initially and angles cannot be accumulated. The circle cannot move with the alidade but its position can be changed between sets of observations so as to have independent readings and distribute any possible errors in markings.

Angles are measured between survey lines. These lines commonly form a *traverse*, a series of interconnected lines. Traverses usually form closed polygons. Closing a traverse in this way provides a geometric check in that the sum of the interior angles adds to (n-2) 180° where n = the number of traverse points.

Theodolite angles are generally measured clockwise because circles usually are graduated that way. If a closed traverse is measured the angles are usually *interior.* The initial sighting to the left for *orientation* is called the *backsight* (BS). The *foresight* (FS) station is the one toward which the theodolite is pointed to measure an angle or determine a new direction.

Sample field data for horizontal angle measurement are shown in Figure 3.16. Also shown with each data set are computations for the mean angles. The purpose of the illustration is to present a study exercise for better understanding of the angle measuring process in each case. The sample data forms are not necessarily those used by all agencies and are somewhat simplified for this illustration. As an aid to studying the direction observations, it must be recognized that there were four settings of the circle, but eight observations of each direction. Angles were computed by taking the differences between directions. Computations shown are only for the first set of observations. The calculations for the other three sets and for the final mean values are left as a student exercise.

Alignment. The second principal function of a theodolite is establishing points on a straight line. For this, the angle measuring features are not needed. The line of sight is used. It is caused to revolve through a vertical plane by rotating the telescope about the horizontal axis which is made horizontal by proper leveling of the plate bubbles and instrument adjustment. Points can thus be set anywhere in the vertical plane through the instrument station by raising or lowering the telescope and aligning the points anywhere along the image of the vertical cross-hair. In practice, a remote station is usually marked with a target and orientation from the instrument made on this point. Then other ground stations can be set "on line" between the instrument and the target station, or beyond the target station, or in the opposite direction by "plunging" or inverting the telescope.

There are many applications of this alignment feature in land, engineering, and other types of surveys.

Vertical angles. In trigonometric leveling, indirect measurements for heights, topographic surveys, some astronomical observations for azimuth and positions, and other uses, either a vertical or zenith angle must be measured. Vertical angle measurement with a theodolite is simpler than horizontal angle measurement. For most purposes, two observations are made—as usual, once with the telescope direct and once with it reversed. The second reading is used as a check on the first and the two, when averaged, provide a means to compensate for circle graduation or index errors. The horizontal cross-hair, rather than the vertical cross-hair, provides the line of sight. Figure 3.17 shows a set of zenith angle measurements and the averaging of these two angles.

Leveling. Although not one of its principal functions, a theodolite can be used to determine elevation differences. The line of sight is established in a horizontal plane by proper leveling of the plate bubbles, setting the zenith angle to read 90°, and proper adjustment of the instrument, after which elevation differences can be observed using differential leveling procedures (see Section 3-5). The theodolite is not normally used for this purpose since level instruments are generally more precise, lighter, less expensive, and faster to operate.

Double Angle Procedure

⊼	Sight	D	R	Mean	1′ Transit
A	B	0° 00′	32° 43′		
				32° 43.5′	
	C	32° 43′	65° 27′		

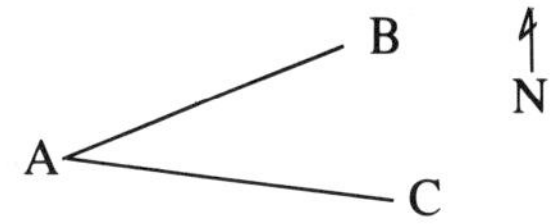

Figure 3.16 (a) Double Angle Procedure

Angles by Repetition

⊼	Reps	Read	Angle	Repeating Theodolite
D	0	0° 00′ 00″		
D	1	47° 10′ 00″		
D	3	141° 30′ 20″		
R	6	283° 00′ 25″		

$$\text{Angle} = \frac{283^\circ\,00'\,25''}{6} = 47^\circ\,10'\,04''$$

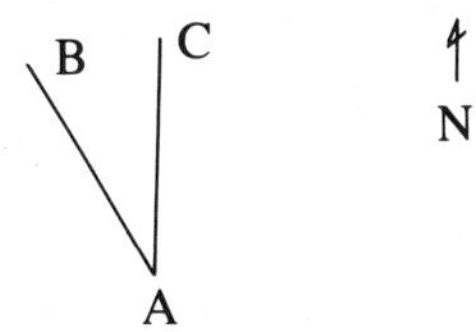

Figure 3.16 (b) Angles by Repetition

Observing Angles, Direction Method

	D			R				D			R	Mean
Sta.	°	′	″	°	′	″	Angle Point	°	′	″	″	″
1	0	01	33.0	180	01	27.0	1-2	106	22	49.5	48.9	49.2
2	106	24	22.5	286	24	15.9	2-3	93	17	36.4	38.1	37.2
3	199	41	58.9	19	41	54.0	3-1	160	19	34.1	33.0	33.6
								360	00	00.0		120.0
1	45	02	23.8	225	02	24.9						
2	151	25	24.5	331	25	18.4						
3	244	42	57.0	64	42	54.2						
1	90	02	16.4	270	02	07.5						
2	196	25	08.4	16	25	02.5						
3	289	42	43.9	109	42	40.2						
1	135	01	53.0	315	01	41.5						
2	241	24	37.8	61	24	34.2						
3	334	42	22.5	154	42	15.1						

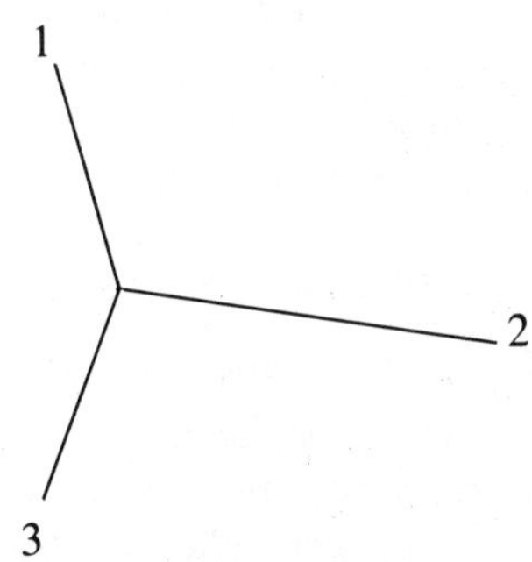

Figure 3.16 (c) Observing Angles, Direction Method

Zenith Angle Readings

Telescope Position	Reading	Correction	Adjusted Value
Direct	86° 14.4′	−0.8′	86° 13.6′
Inverted	273° 47.2′	−0.8′	273° 46.4′
	359° 61.6′	−1.6′	359° 60.0′

Error = (359° 61.6′) −360° = + 1.6

Figure 3.17 Zenith Angle Readings

Stadia. Distance up to a few hundred feet can be measured to accuracies of approximately one or two feet employing stadia hairs built into the telescope of most theodolites. This will be mentioned under the heading of distance measurements later in this chapter and analyzed in Chapter 5.

Targets

Angle, direction, or alignment operations nearly always require the use of some sort of target. If ground stations can be sighted directly, targets are not necessary, but terrain and range usually prohibit this and something must be used to define a vertical line through the station. For horizontal alignment, anything which is straight and which can be centered over the station can be used. *Range poles* are the most common instruments. They vary in length from about 5 feet to over 20 feet. They are usually round with a pointed end, from 0.5 to 1.5 inch diameter, and made of metal or wood. Such instruments are held vertically by a rodman in order to define a vertical line for sighting by the instrument operator. A range pole in use is shown in Figure 3.18.

For short ranges, plumb bobs can be centered over stations and the string sighted as a target. Pencils and taping pins can also be used for short sight distances. The ideal target size is a diameter that is not much greater than two or three times the cross hair width as viewed through the instrument. This ideal size varies with the magnification of the telescope, target distance, and atmospheric conditions.

For precise work, geodetic targets, mounted on tripods, are used. These have a distinct pattern that can be seen from long distances. They have provision for illumination in order to make night observations or perform surveys in mines or tunnels. Fig. 3.19 shows such a target.

3-4 DISTANCE MEASURING INSTRUMENTS

Tapes and Similar Instruments

One group of distance measuring instruments consists of tapes, chains, lines, ropes, cords, rules, rods, and other calibrated linear devices which can be laid end for end to accumulate a distance between points. The ancient surveyors used ropes treated in wax which were calibrated in cubits or other ancient units. Lines and cords were also used for many centuries. Rods, fixed in length (such as 16.5 feet,

Figure 3.18 Range Pole in Use

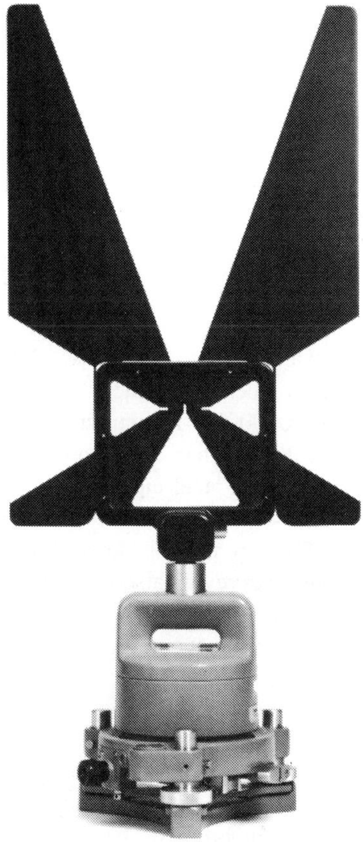

Figure 3.19 Geodetic Target (Courtesy of Wild Heerbrugg Instruments, Inc., Farmingdale, N.Y.)

for instance) have been used in Renaissance and early modern times.

Edmund Gunter's *chain* of 66 feet in length was somewhat of a breakthrough in direct linear distance instrumentation. It consisted of 100 metal links, each 0.66 feet long. There was a tag every ten links to facilitate reading fractional lengths of the chain. An *engineer's chain* has been used in the past which is 100 feet long. Except for its length, the engineer's chain is similar to the Gunter's chain.

Such instruments as ropes, lines, cords, rods, and chains are practically obsolete, although precisely calibrated wires are sometimes used for certain measurements. Tapes are the most commonly used direct distance measuring instruments. They are graduated in a choice of units—generally feet, chains, or meters. Tapes are made of various materials, including steel, invar (a nickle-steel alloy), non-metallic (woven tapes made of synthetic yarns), and metallic woven tapes (having metallic strands woven in with the yarn).

The traditional surveyor's tape in the United States has been the 100 foot steel tape. The 100 foot steel tape will likely be replaced by the metric steel tape as we progress toward implementation of the metric system. Metric steel tapes are available in 20, 25, 30, 50, 100 meter and other lengths.

Typical characteristics of a steel tape are as follows:

thickness – 0.016 to 0.025 inches (0.041 to 0.064 cm)

width – 0.25 to 0.375 inches (0.64 to 0.95 cm)

weight (mass) – 2 to 3 lbs (0.9 to 1.4 Kg) (100 ft. tape)

coefficient of thermal expansion – 0.00000645/°F (0.0000116/°C)

modulus of elasticity – 29,000,000 lbs/in^2 (~2,050,000 Kg/cm^2)

Invar tapes have been used for precise control surveying. Their advantage over steel tapes is that the coefficient of thermal expansion is about 1/30 that of steel (0.00000022/°F), thus errors due to temperature effects are minimized. Invar is more brittle than steel and breaks more easily. Invar tapes are also more expensive than steel tapes.

The woven tapes (sometimes called "cloth" tapes) are not made for accurate work. They are made in somewhat shorter lengths than steel tapes (perhaps 25 to 100 feet). Surveyors also use simple 6-foot folding rules or pocket reel tapes for short measurements.

Pictures of several types of tapes and rules are shown in Figs. 3.20 to 3.22. It can be seen that the graduations and method of marking the tapes varies. Some tapes have an extra foot beyond the zero mark which is graduated in tenths and possibly hundredths. This is called an *add tape*. If the foot between the zero and 1-foot mark is graduated and there is no extra foot added, the tape is called a *cut tape*. If the popular "add" feature is used on a metric tape, the section added at the zero end will probably be 1-decimeter in length and be subdivided into millimeters, with the full tape divided into decimeters.

Various accessories are used in taping with a steel tape. A *tape clamp* is used by the rear tapeman for gripping the tape as the head tapeman applies tension. A *tension handle* (also called spring balance) is usually attached at the head of the tape for measuring the tension applied. A *tape thermometer* is used for reading the air temperature in the vicinity of the tape. *Taping pins* are used for marking each tape length and for keeping a tally of the tape lengths. These and other tape accessories are shown in Fig. 3.23. A *range pole* is used for placing behind the forward station so that a direction can be defined for alignment purposes. *Plumb bobs* are used for marking the tape ends on the ground when the end of the tape must be suspended.

Taping Procedures

Taping with a steel tape is performed by laying the tape end for end the required number of times. Successive tape lengths are marked with taping pins or other methods. Each pin set by the head tapeman is removed by the rear tapeman after the head tapeman has set the next one. Alignment is accomplished by setting a target, such as a range pole at the forward point. Hand signals from the rear tapeman or a theodolite operator at the beginning point are used to keep the head tapeman in line.

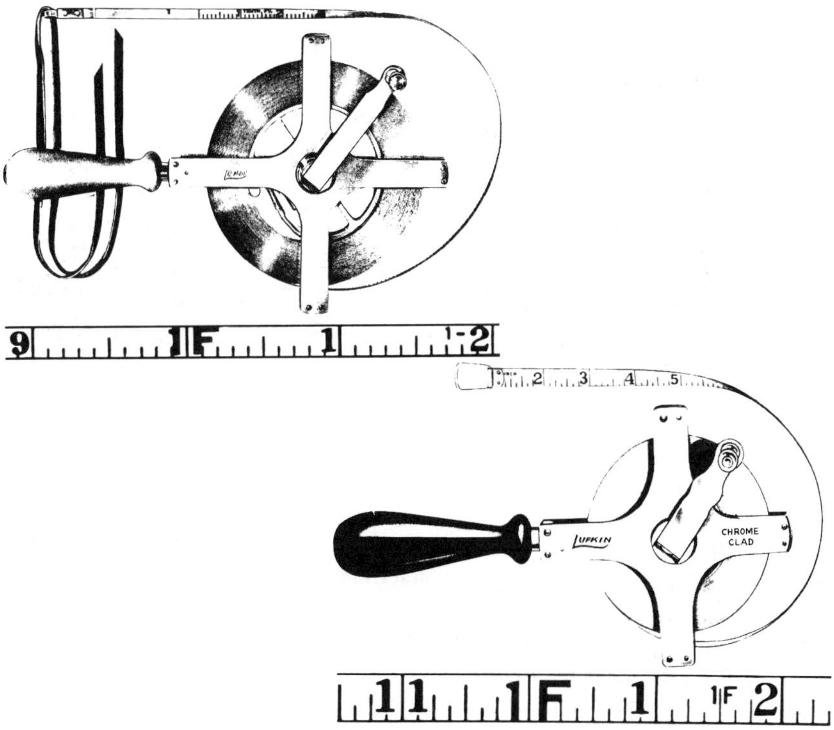

Figure 3.20 Steel Tapes (Courtesy of Lufkin ® Company (The Cooper Group), Raleigh, North Carolina.)

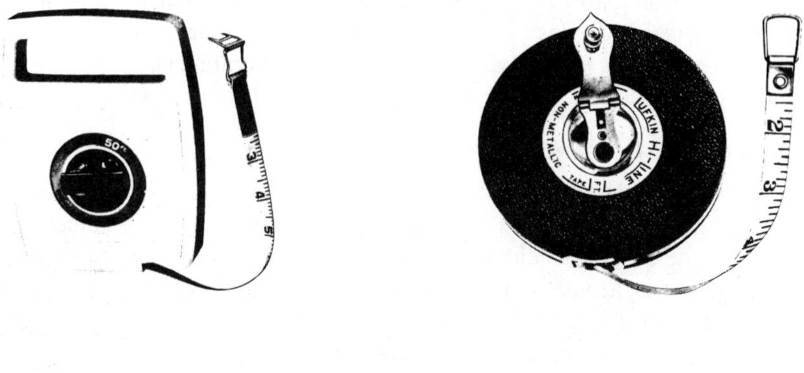

Figure 3.21 Non-Metallic Tapes (Courtesy of Lufkin ® Company (The Cooper Group), Raleigh, North Carolina.)

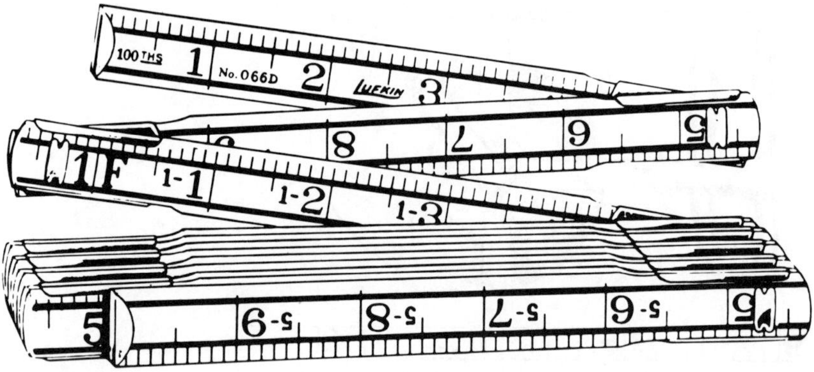

Figure 3.22 Folding Rule (Courtesy of Lufkin ® Company (The Cooper Group), Raleigh, North Carolina.)

As should be recalled from Chapter 1, the distance desired is horizontal distance. This presents a problem in that terrain is rarely flat (at 0% slope) and it is irregular, having many hills and valleys both large and small. Distances can be reduced to horizontal after measuring the slope of a tape with the aid of a hand level, clinometer, or theodolite. Horizontal distances can be measured directly by holding the tape ends at the same elevation, but then the marking on the end which is suspended must be transferred to the ground using a plumb bob and corrections must be made for the effect of sag of the tape.

The tape length is affected by temperature and tension. It stretches or expands with added tension or higher temperature. Therefore, in order to control accuracy, measurements or estimates of these variables must be made as work progresses. The tape is normally calibrated under controlled conditions of temperature and tension so that any errors in manufacture are discovered. The effects of temperature and tension follow known physical laws and thus are predictable, such effects being classical examples of systematic errors in surveying.

The equations for applying corrections and controlling taping errors will be introduced in Chapter 5 where they will be analyzed and thoroughly discussed.

Tacheometric Instruments

The word tacheometry derives from the Greek words, ''taklus'' which means swift or rapid, and ''metron'' meaning measurement. Any measurement which can be made quickly would literally be tacheometric. Commonly accepted categorization of tacheometric instruments would probably exclude electronic measurements even though they are made rapidly. The simplest literal tacheometric measurement would be estimation. One can make an educated judgment (not merely a guess) of a distance, angle, direction, or height based on experience and familiarization with relative unit values. For example, distance might be estimated by observing a person's relative size with respect to surrounding objects. *Estimation* is a valid measurement method for certain rough measurements. Most modern

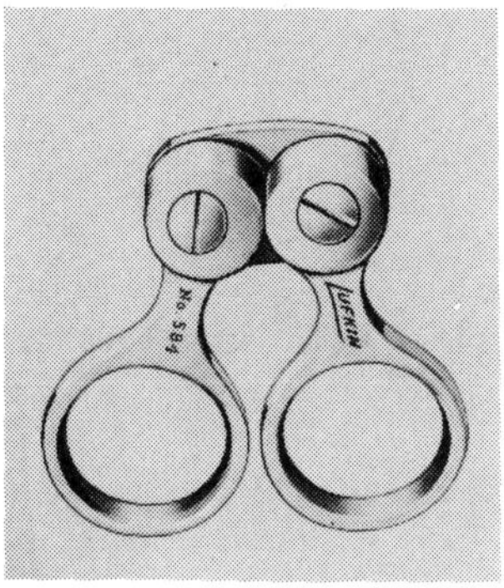

TAPE CLAMP

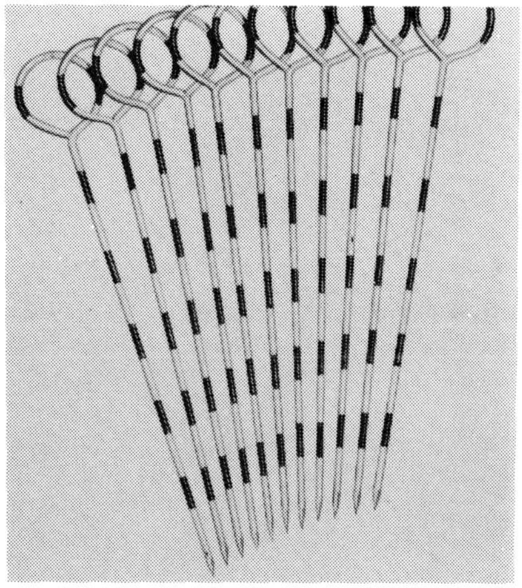

TAPING PINS

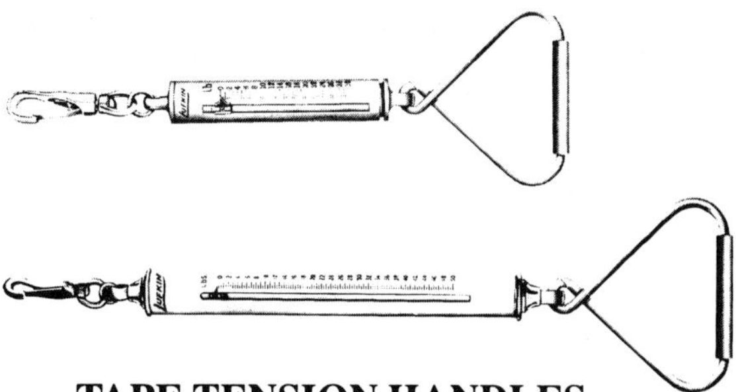

TAPE TENSION HANDLES

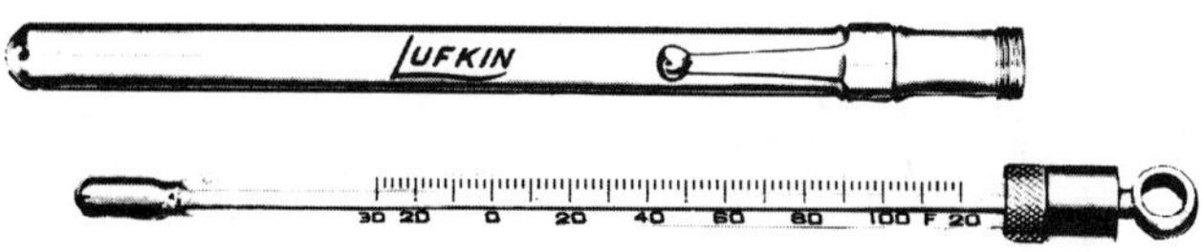

POCKET THERMOMETER

Figure 3.23 Tape Accessories (Courtesy of Lufkin ® Company (The Cooper Group), Raleigh, North Carolina.)

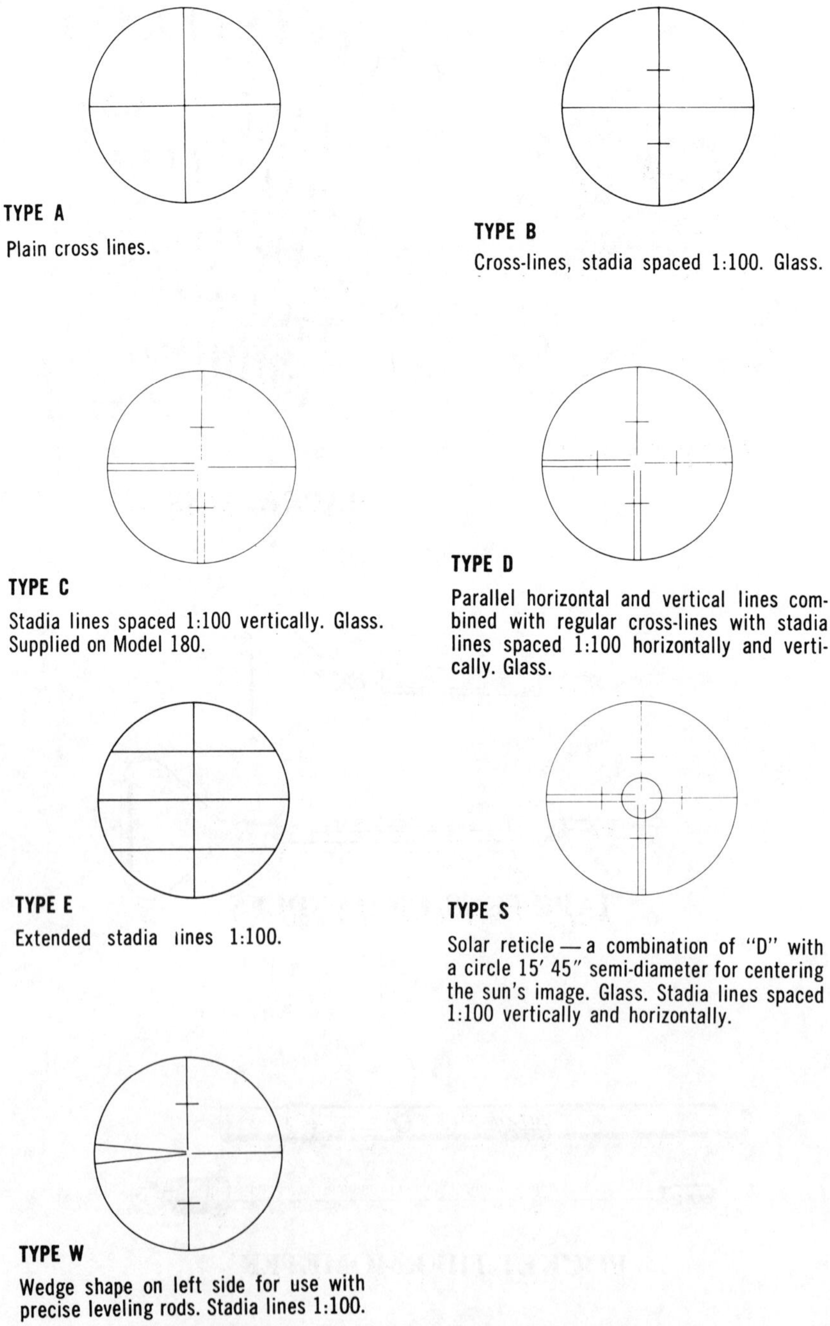

Figure 3.24 Reticles (Courtesy of The Lietz Company, Overland Park, Kansas.)

textbooks would list tacheometric instruments as falling into two groups: (1) those employing a *stadia* principle and (2) those employing a *subtended angle* principle. Both horizontal and vertical distances can be determined with most instruments which employ these principles. The stadia principle is used in various ways resulting in several variations of instruments. This section will be confined to a listing of the common instruments and their general use and application. The various formulas involved will be derived and applied in Chapter 5.

Stadia instruments measure distance using the principle that the intercept or difference in reading between two fixed cross hairs is directly proportional to the distance between the target (level rod or stadia rod) and the telescope. Most theodolites have an upper and lower stadia hair in addition to the center horizontal hair. These are fixed so that there is a constant multiplier for converting the rod intercept to distance. When the vertical angle to a specific rod reading is measured along with the rod intercept, vertical elevation differences can be computed. The most commonly used stadia instruments are ordinary theodolites. See Fig. 3.24 for various stadia reticles employed.

Some stadia instruments are *self reducing* which means that distances can be read more or less directly without the need for calculations. This is accomplished by special arrangement of curved lines in the reticle.

From the other group of tacheometric instruments, the *subtense bar* is the

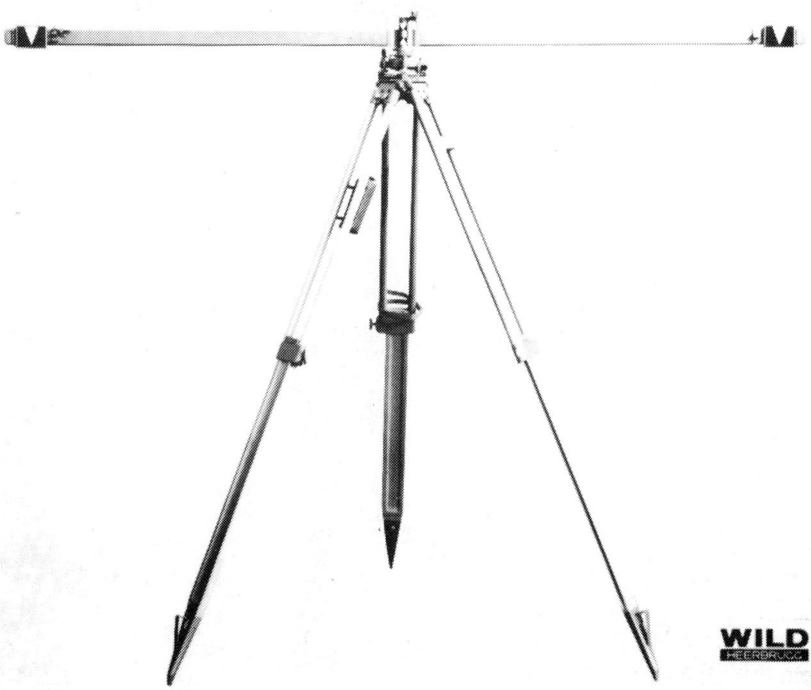

Figure 3.25 Subtense Bar (Courtesy of Wild Heerbrugg Instruments, Inc., Farmingdale, N.Y.)

most commonly used. See Figure 3.25. The bar maintains a known length between two targets near its ends. It is erected perpendicular to the survey line and centered over a survey station. The subtended angle is measured between the targets using a theodolite. The horizontal distance is computed from the subtended angle using trigonometry.

Figure 3.26 Plane Table and Alidade (Courtesy of Wild Heerbrugg Instruments, Inc., Farmingdale, N.Y.)

A subtense bar is generally made of invar which has a very low coefficient of thermal expansion. It has a base for mounting on a tripod and an optical plummet for centering over the station. This base for the Wild model shown in Fig. 3.25 is a tribrach which provides for interchangability between the theodolite and subtense bar. It has a sighting scope at its center for aligning the bar normal to the line of sight. The two targets at the ends of the bar can be illuminated using a battery attachment for night observations. A typical subtense bar is 2 meters between targets, although they are sometimes made longer.

The field procedures involve centering the theodolite over one station and the subtense bar over the other. Next the bar is aligned perpendicular to the line. The angle between the targets is then measured. The number of measurements made depends on the precision desired. The mean angle is used to compute the distance. Calculations will be explained in Chapter 5.

Other tacheometric instruments such as the *distance wedge*, the *reduction tacheometer*, and other designs are not discussed herein. Those discussed and illustrated represent the basic instruments used.

Special mention should be made, however, of the *alidade* used in conjunction with the *plane table*. This alidade is a telescope with a vertical motion and stadia hairs as on any transit or theodolite. It is mounted on a flat straight edge rather than on a base plate with foot screws as on a theodolite. It sets on top of the plane table and is used to graphically define directions and plot distances. An alidade and plane table are illustrated in Fig. 3.26.

Electronic Distance Measuring Instruments

This generation of surveyors has seen a revolution in the methods of measuring distances, particularly long distances over rugged terrain. The first instrument for measuring distance electronically was developed by Dr. Bergstrand in Sweden in 1946. It was called a geodimeter and it employed a modulated light beam for determining distance. This instrument went through various stages of development and was used in control surveying in the early 1950's. It was followed by an instrument called a tellurometer, developed by T.L. Wadley of South Africa in 1956. The tellurometer employed modulated microwaves for determining distance. These two instruments represent the basic principles still used for making measurements electronically—employment of either a light beam or a microwave.

Electronic distance instruments were not used very much for most surveying work until the 1960's. The early models were expensive, quite heavy and bulky, required generators or heavy batteries for power, and their use and maintenance generally required some skill beyond that of the ordinary surveying technician. But, their advantages of measuring long distances (as much as 10 miles or more) over any type of terrain was sufficient to spur research and development. With modern circuitry and other developments, the equipment has gradually become less bulky and expensive. But, the range of the lighter models is much shorter than the earlier ones. There is hardly a progressive surveying firm today that does not employ electronic distance equipment (EDM) to some extent.

The light wave type of instruments employ frequencies in the visible light wavelength spectrum and the infrared range. The equipment consists of an electrical unit and a retro prism. The electrical unit transmits the light to the prism which reflects it back to the unit for analysis of the double slope distance between the two points. The returned light is converted to electrical current by a photo tube. The wavelength of the light is known and by a method of phase comparison, the partial wavelength can be determined.

$$S = \tfrac{1}{2}(n\lambda + d)$$

where

S = the slope distance between the measuring unit and the reflector,

n = the number of full wavelengths,

λ = the wavelength,

d = the partial wavelength.

The number, n, is resolved by using more than one frequency and solving the equations simultaneously. Actually, this is all done almost automatically in modern instruments. Older instruments required taking raw readings of the phase change on a null meter and converting the readings to distance by computation. The wavelength, λ, is determined by

$$\lambda = \frac{\nu_a}{f_m}$$

where

ν_a = the velocity of the light through the atmosphere

f_m = the modulating frequency, in hertz.

The principles discussed above can be illustrated as follows:

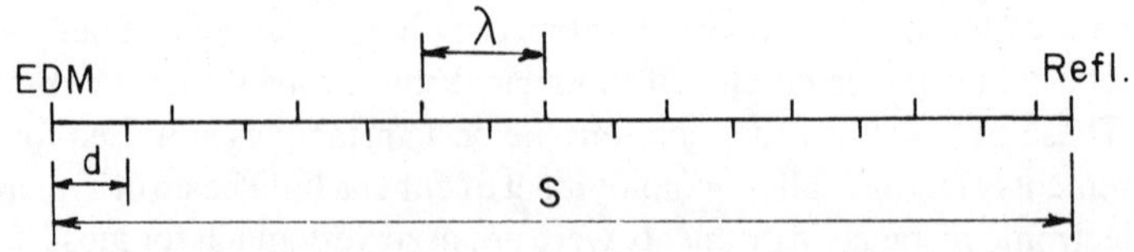

The microwave type of instrument employs two electrical units, one at each end of the line being measured. Each unit has a transmitter, receiver, antenna, measurement circuitry, and built-in communication system. The principle of measurement is similar to that of the light wave type. The frequencies used in EDM are as shown in Table 3.1.

The various models of EDM available change constantly. New features are continually added and so it is quite impractical to list and discuss every model. There were over twenty popular models available in the late 1970's with prices ranging from approximately $4,000 to $10,000 for the short range (up to about

TABLE 3.1

Frequencies Used for EDM — Visible Light to Microwave

λ, meters	f_m, cyc/sec		Nature	Property
3000 - 100	10 - 300 kc/sec		very long radio waves	Travel along the curvature of earth
1000 - 100	300 - 3000 kc/sec		medium	
100 - 1	3 - 300 mc/sec			
1 - 0.01	300 - 30,000 mc/sec	↑	microwave	Wave starts traveling in straight line and
$0.01 - 10^{-6}$		EDM	infrared	penetration through denser mediums
$0.4\mu m - 0.8\mu m$ (microns)		↓	visible light	increases.
$10^{-8}cm - 10^{-10}cm$			X-rays ↓ cosmic rays	Will penetrate lead

3,000 meters) instruments and generally over $10,000 for those with ranges up to 25 kilometers. Very long range instruments (60 to 80 kilometers) may cost up to $50,000. Accuracies vary from perhaps ±2mm. to ± 2cm. plus another smaller error which is directly proportional to distance.

For current information on EDM, it is suggested that the reader request brochures from suppliers and also consult the references on the subject listed at the end of the chapter.

3-5 ELEVATION INSTRUMENTS

Direct Leveling Instruments

Levels are some of the simplest instruments used by surveyors. In principle, surveying levels differ little from carpenter's levels. If one can imagine the flat edge of the carpenter's level representing a line of sight through a surveyor's telescope, the construction of a surveyor's level is easily seen. The simplest surveyor's level has a level vial the axis of which is made to be parallel to the telescope line of sight. Similarly, a carpenter's level is made so that the bubble axis is parallel to the flat, straight edge. When the bubble of either instrument is centered, the edge (line of sight) is horizontal.

The self-leveling or *automatic level* is probably the most widely used level instrument. It has a compensator which swings free when a circular bubble on the base of the instrument is centered. When the compensator comes to rest (almost instantaneously), the line of sight is horizontal, (assuming it is in adjustment) even with slight changes in the circular bubble centering. The advantages of an automatic level are the speed of operation and added precision resulting from the ever

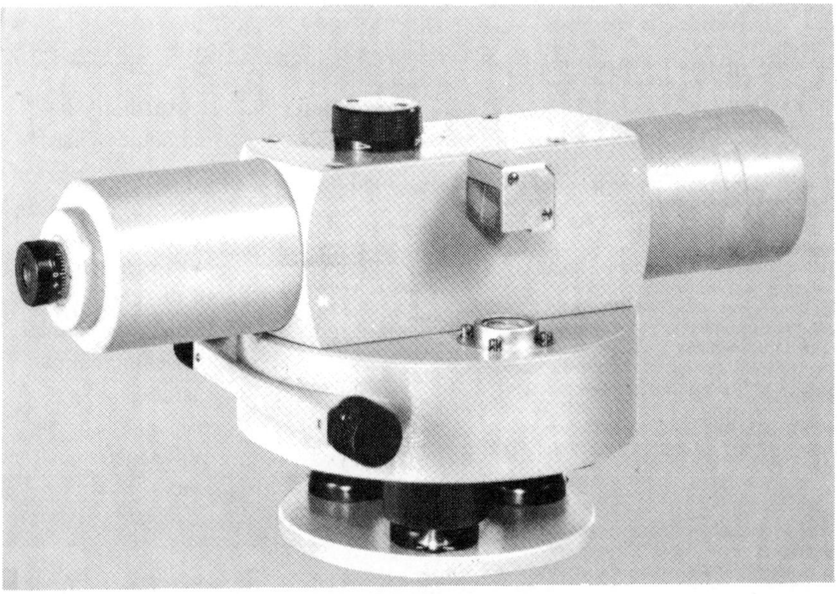

Figure 3.27 Automatic Level (Zeiss level used with permission of Keuffel and Esser Company, Morristown, New Jersey.)

constant horizontal line of sight. The features of an automatic level are illustrated in Figure 3.27.

Tilting levels employ the split image bubble described in Section 3-2. The telescope and bubble can be tilted slightly about a horizontal axis. The instrument shown in Figure 3.28 is leveled by first centering the circular bubble using the three foot screws, then centering the more sensitive longitudinal bubble using the knob located under the telescope. Tilting levels are somewhat slower to use than automatic levels but are faster than the dumpy and wye levels. They are generally priced a little lower than automatics.

One of the earliest levels, called a *wye level* was invented in the eighteenth century. The *dumpy level* was developed during the nineteenth century. Both are still used to some extent today, although they are rapidly being replaced by tilting and automatic levels. The wye level has a telescope that rests in a Y-shaped support. It can be detached and pointed in the opposite direction. This feature aids in performing certain adjustments to the instruments. Leveling is done using foot screws attached to the base of the instrument. The longitudinal bubble is attached beneath the telescope. The dumpy level is constructed similar to the wye level except that the telescope cannot be detached. Both the wye and dumpy levels are much slower and less precise than either the automatic or tilting level due to a general lack of sensitivity in the leveling system and difficulty in assuring that the

Figure 3.27 B (NA2 level and compensator courtesy of Wild Heerbrugg Instruments, Inc., Farmingdale, N.Y.)

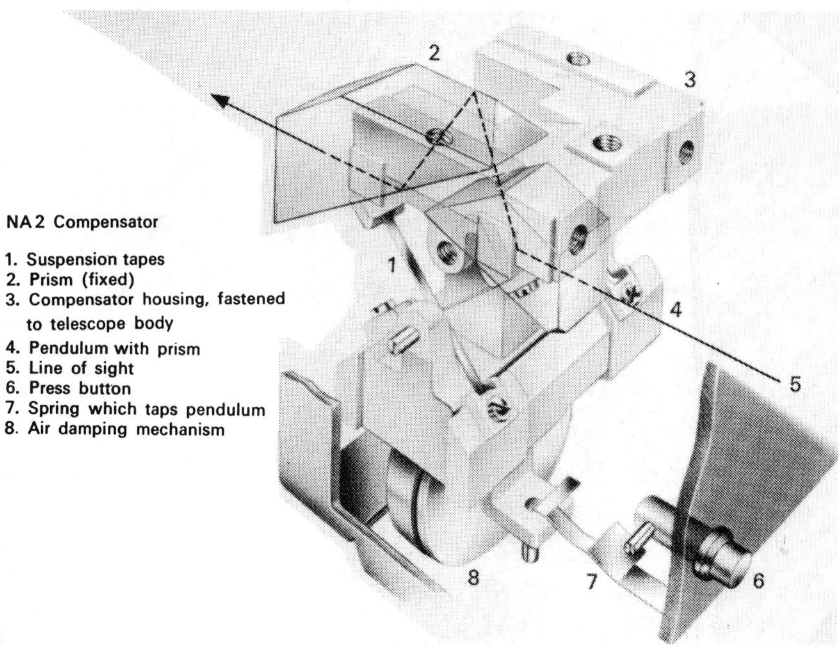

Figure 3.28 Tilting Level Courtesy of The Lietz Company, Overland Park, Kansas.)

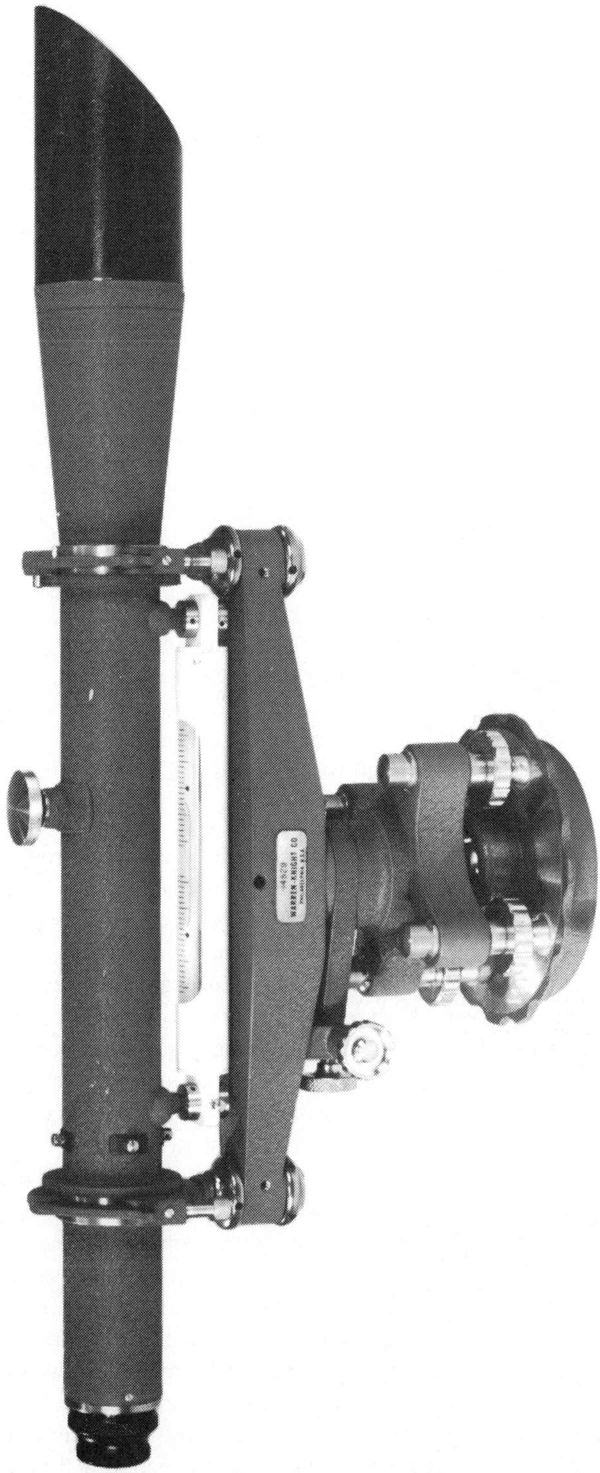

Figure 3.29 The Wye Level (Courtesy of Warren-Knight Instrument Co. Philadelphia, Pa.)

bubble is truly centered at the moment of observation (see Figs. 3.29 and 3.30).

A *hand level* is a variation of the above, except that it usually has no magnification, is hand held rather than tripod mounted, and is much less precise. Hand levels are used for rough leveling and for determining the horizontal in taping. A hand level with carrying case is shown in Fig. 3.31. This instrument is approximately six inches in length.

Level instruments (excluding hand levels) have some common components: telescope with horizontal cross-hair, a leveling mechanism (bubble, compensator, etc.) foot screws for leveling, base plate, and tripod.

Barometric Leveling Instruments

A barometer is an instrument which measures air pressure. An altimeter is an instrument which measures elevation by changes in air pressure. Altimeters are the instruments used for barometric leveling. They are calibrated in scales reading to ± 2 ft. to ± 10 ft. Air pressure is affected by elevation differences as well as temperature, humidity, and changing weather so that methods must be used to control these effects. Altimeters are not commonly used in most surveying operations.

Trigonometric Leveling Instruments

Any theodolite, along with a distance measuring instrument can be used for trigonometric leveling. All that is required is the determination of a vertical angle and horizontal or slope distance from which the elevation difference is computed by trigonometry. Methods and calculations for this and effects of curvature of the earth and atmospheric refraction will be discussed in Chapter 7.

Level Rods and Accessories

Direct and trigonometric leveling requires the use of level rods which are nothing more than vertically held scales. They are used to measure the height of a line of sight above the ground station. Leveling rods are made in various lengths from about 7 feet to 15 feet or more. Rods are usually made to extend in two or more sections, each section being from 4 to 7 feet in length. Precise rods are made in only one section, however. They are made either with graduations to hundredths of feet or centimeters. Distinctive marking designs are employed to facilitate reading. Level rods for engineering and land surveying work usually have all markings directly on the rod face. Precise rods have the graduations on invar strips with the numbers on the wood face. Level rods are shown in Figure 3.32. The rod on the left is a precision geodetic rod. The other two are ordinary rods used in land and engineering surveying. One is graduated in feet, the other in meters.

Level rods are generally read directly by observing the intersection of the horizontal cross hair image on the rod, which is held vertically on a survey station. Such rods are sometimes called *self reading*. When sight distances are greater than about 400 - 500 feet or when the markings are not distinct, readings are often made with the aid of a *rod target*. The target is attached to the rod and has its own distinct pattern which can be seen much clearer than the rod graduations. It is raised or lowered along the face of the rod by the rodman until the instrument operator signals that it is aligned. The rodman then observes the reading at the index mark.

Figure 3.30 The Dumpy Level (Courtesy of The Lietz Company, Overland Park Kansas.)

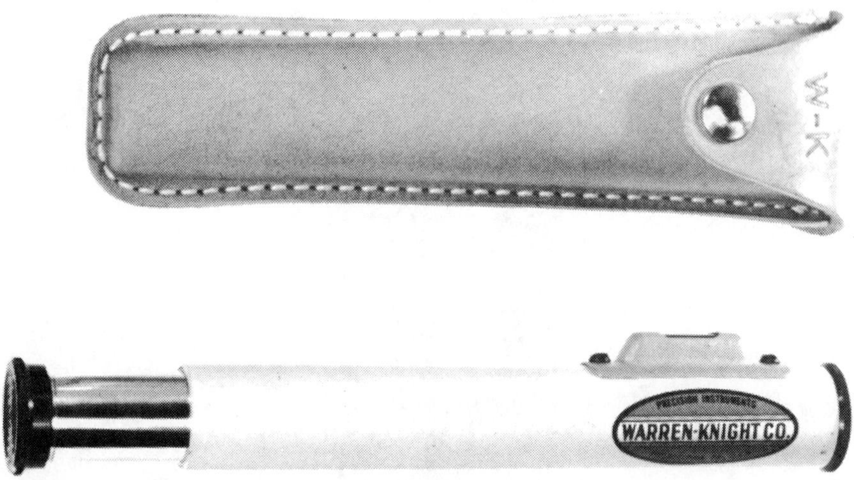

Figure 3.31 Hand Level (Courtesy of Warren-Knight Instrument Co. Philadelphia, Pa.)

Targets are equipped with verniers to enable reading to a tenth of the smallest rod division. A rod target is shown in Fig. 3.33.

Level rods must be held vertically so that the true vertical distance above the rod station is observed. Precise level rods usually have a circular bubble attached to them for plumbing. Ordinary rods do not usually have such a bubble attached but a detached bubble can be used in a similar manner (See Fig. 3.34).

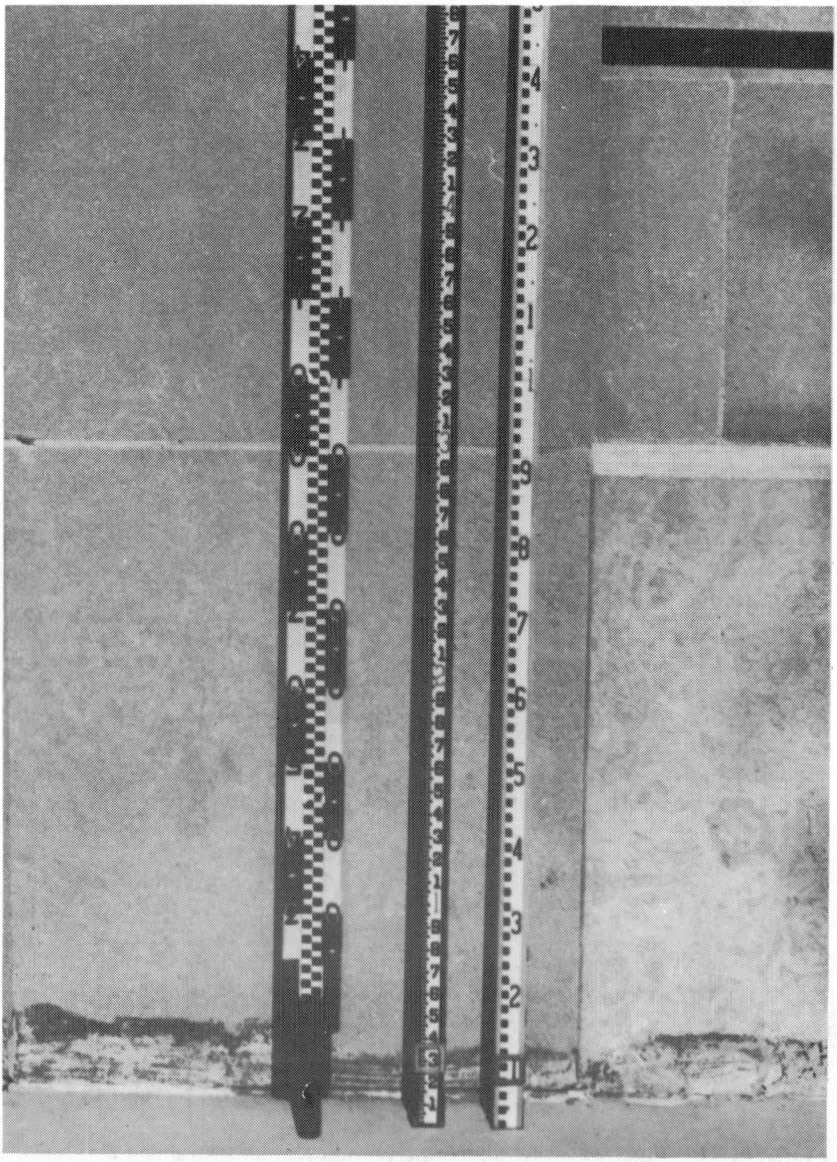

Figure 3.32 Level Rods

Figure 3.33 Rod Target Attached To Level Rod

Leveling Procedures

Elevation differences are generally determined using a procedure called *differential leveling* (also called spirit leveling or geometric leveling). Either an automatic, tilting, dumpy, wye, or hand level is used with a level rod.

The procedures can be surmised from the concepts and definitions given in Section 1-6. Figure 3.35 illustrates the basic principle. An instrument is set up half way between two points whose elevation difference is desired. The instrument is "leveled" –that is, the line of sight made to be horizontal. Rod readings (r_A, r_B) are taken on the two points, a level rod being held vertically on each point. The difference in elevation is

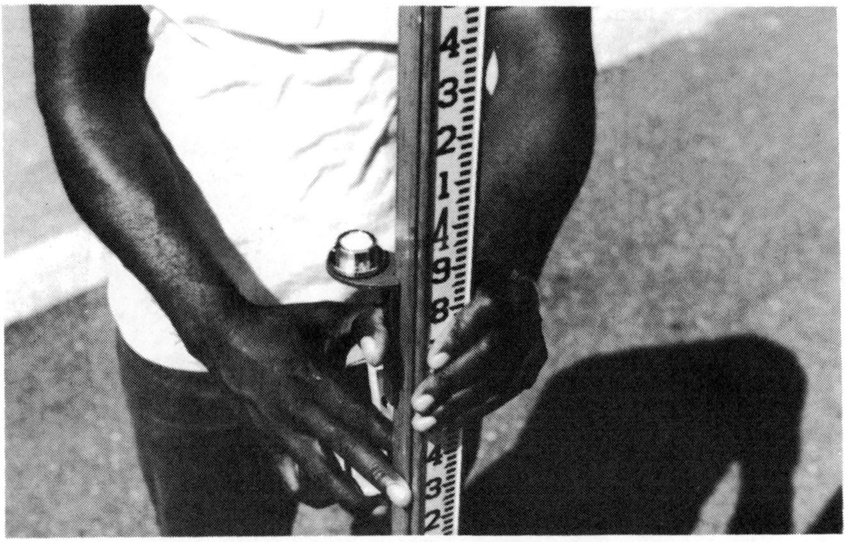

Figure 3.34 Circular Rod Bubble Being Used To Plumb Rod

$$\Delta\, \text{Elev}_{A-B} = r_A - r_B \tag{3.3}$$

Actually, the line of sight does not point exactly horizontal due to effects of atmospheric refraction. These effects are quite small for normal sight distances of approximately 100 meters or less and cancel in the process anyway if sight distances to the backsight and foresight stations (A and B in Fig. 3.35) are equal. In a similar manner, effects of the earth's curvature are cancelled when sight distances are equalized. More will be discussed about these effects in Chapter 7.

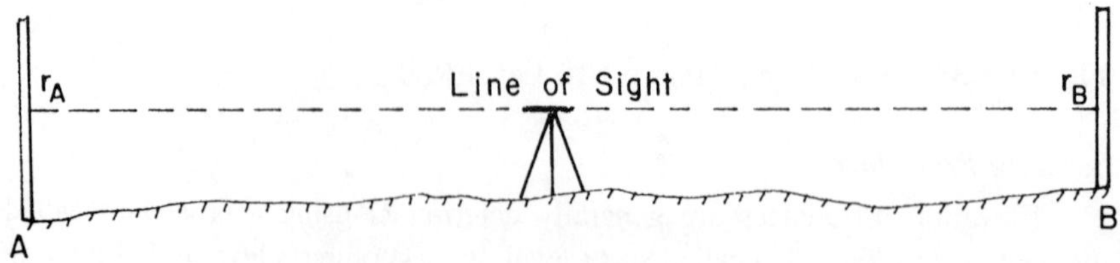

Figure 3.35 Differential Leveling Principle

The above procedure is adequate if the instrument to rod distance permits clear reading of the rod at both stations and if the elevation difference is small enough so that horizontal sightings are neither below nor above the rod at one station or the other. If distances are too long or elevation differences too great, then more than one instrument set up must be made. The following set of definitions and Figs. 3.36 - 3.37 explain the data gathering procedure.

A *backsight* (BS) is the vertical distance between the line of sight and a point of known or assumed elevation, or the reading taken on such a point.

A *foresight* (FS) is the vertical distance between the line of sight and a point whose elevation is desired, or the reading taken on such a point.

The *height of instrument* (HI) is the vertical distance of the line of sight above a datum which is computed by adding the elevation of a point to the BS on that point.

A *benchmark* (BM) is a point of known elevation.

A *turning point* (TP) is a point whose elevation has been determined by a FS. The elevation of the TP is determined by subtracting the FS from the HI.

If absolute elevations above a datum are not needed, differential leveling can start and end on any two points whose elevation difference is desired. If an elevation is needed on one point, the starting point must be known in elevation. The field procedures are the same in either case.

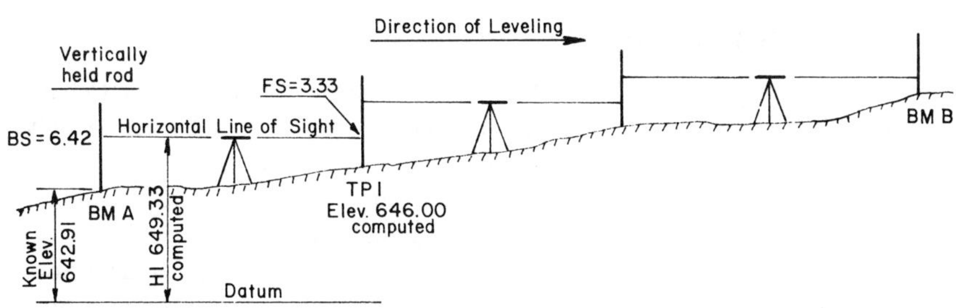

Figure 3.36 Differential Leveling Procedures

Station	BS	HI	FS	Elev.	Adj. Elev.
BM A	6.42	649.33		642.91	
	7.01	653.01	3.33	646.00	
	5.12	655.63	2.50	650.51	
BM B	0.53	654.67	1.49	654.14	
	2.95	650.70	6.92	647.75	
	3.80	649.20	5.30	645.40	
BM A			6.23	642.97	

Figure 3.37 Differential Leveling Notes

Figure 3.36 illustrates the field procedures for a leveling operation from BMA to BMB. It will be noted that Fig. 3.37 also contains the readings for the second half of the circuit, closing back to BMA. A circuit must always be closed on a point of *known* elevation if the established elevations are to have an adequate check for errors and mistakes. It is not necessary to close on the same bench mark, but a closure must be made to a benchmark on the same datum. Even an elevation difference measured without reference to a bench mark must be checked by *closing the circuit* to the starting point.

It should be obvious from a study of Figs. 3.36 and 3.37 that the rod is held on the same spot for both the foresight and backsight on that particular turning point. The level instrument is moved to new locations as necessitated by range of the telescope and terrain after which it is releveled prior to taking rod readings.

In the example, the computed elevation difference of BMA is different than the starting elevation by +0.06 ft. Assuming that there are no mistakes and that systematic errors are compensated, this is an *error of closure* representing an accumulation of random errors. Dealing with this closure and errors in leveling will be included in Chapter 7.

3-6 CALIBRATION OF EQUIPMENT

As has been stated in other sections of this text, it is quite possible to have good precision in measurements without good accuracy. The latter occurs only when systematic errors have been detected and removed or compensated. One of the most common oversights made by practicing surveyors is ignoring the calibration error. For example, some surveyors assume that a steel tape is always manufactured to be its nominal length at some arbitrary temperature and tension. The accuracy of electronic equipment often goes unquestioned based on reasoning that since it is so expensive and precise, it surely must also be accurate. Such assumptions are made only by uninformed individuals. There is nearly always a *calibration error* or *index error* unique to any individual instrument. Such instrumental errors can often be corrected by adjusting or ''calibrating'' the instrument. Sometimes they cannot and must therefore be treated as systematic or constant errors and corrected by computation.

Methods of testing and correcting for calibration errors will be discussed as part of the subject matter in each section of Chapters 5-7.

Problems - Chapter 3

3.1 Sketch the line of sight through a surveyor's telescope, being sure to indicate points defining this line.

3.2 What is the purpose of the eyepiece in a surveyor's telescope?

3.3 Explain the difference between an external focussing and an internal focussing telescope.

3.4 What optical quality determines the pointing accuracy of a telescope?

3.5 What is the most common cause of parallax in focussing a surveyor's telescope?

3.6 What is the radius of a bubble tube with 20 seconds sensitivity and 2 mm markings?

3.7 If the bubble tube of Problem 3.6 is centered ±½division, what is the random error in the direction of pointing of its axis due to this centering error?

3.8 What is an important advantage of an optical plummet over a plumb bob?

3.9 If a circular scale is divided into 20-minute portions and the accompanying vernier has 40 divisions, how closely can this system be read?

3.10 Name some advantages of optical theodolites over vernier theodolites.

3.11 To what do the terms "face left" and "face right" refer?

3.12 What is the purpose of the upper motion on a repeating theodolite?

3.13 What is "orientation" as regards angle measurement with a theodolite?

3.14 In order for both the direct and inverted readings of zenith angles to add to exactly 360°, what must the theodolite circle readings be when the telescope is pointing horizontal?

3.15 How much would a common surveyor's steel tape expand due to a temperature increase of 30°F?

3.16 What are the two common types of tacheometric instruments used in surveying?

3.17 Name two advantages of electronic distance instruments over tapes.

3.18 Name and discuss the common types of surveyors' levels.

3.19 If a surveyor's level is set up between two points A and B, and readings taken on a vertically held rod at the two points, what would be read at B if the reading at A was 4.18 feet and it was known that B was 3.62 feet higher than A?

3.20 What is the purpose of calibration of surveying instruments?

Additional Reading References – Chapter 3

1. Blachut, T.J., Chrzanowski, A., and Saastamoinen, J.H.. *Urban Surveying and Mapping*. New York: Springer-Verlag, 1979, Chapter 3 (EDM) and Chapter 5.

2. Brinker, Russell C. and Wolf, Paul R. *Elementary Surveying*. Sixth Ed. New York: Harper and Row, 1977, Chapter 4, 5, 8.

3. *Definitions of Surveying and Associated Terms,* American Congress on Surveying and Mapping, Falls Church, Virginia. 1978.

4. Kiely, Edmund R. *Surveying Instruments: Their History.* Columbus, Ohio: CARBEN Surveying Reprints, 1979.

5. Moffitt, Francis H. and Bouchard, Harry. *Surveying*. Seventh Ed. New York: Harper and Row, 1981, Chapters 2, 3, 6, 7.

6. Mueller, Ivan I. and Ramsayer, Karl H. *Introduction to Surveying*. New York: Frederick Ungar Publishing Company, 1979, Chapter 3.

7. Saxena, Narendra K. "Electro-Optical Short-Range Surveying Instruments," *Journal of the Surveying and Mapping Division,* American Society of Civil Engineers, Oct., 1975.

8. Schmidt, Milton O. and Rayner, William H. *Fundamentals of Surveying*. Second Ed. New York: D. Van Nostrand Co., 1978, Chapters 2, 3, 4.

9. Strasser, Georg "Heinrich Wild 1877/1977" *Wild Reporter* 13.

10. "The Principle of Modern Surveying Instruments" Wild-Heerbrugg Co., Farmingdale, New York, 1960.

11. "The Theodolite and its Application" Wild-Heerbrugg Co., Farmingdale, New York, 1972.

12. Tomlinson, Raymond W. and Burger, Thomas C. *Electronic Distance Measuring Instruments*. American Congress on Surveying and Mapping, Falls Church, Virginia, 1971.

CHAPTER 4

SURVEY MEASUREMENT THEORY

4-1 THEORY AND PRACTICE

The word "theory" itself seems to almost frighten some surveyors. Some, who lack formal education in measurement theory feel that it is "beyond them." The apprehension about measurement theory is unnecessary. A thorough background in statistics or differential equations is not required to relate measurement theory to survey practice. Many of the refinements with which statiticians and mathematicians concern themselves are not necessary as applied to analysis of survey measurements. However, measurement *is* theoretical. Since exact results remain unknown in any measurement, the surveyor must analyze and estimate the degree of uncertainty. Thus, the surveyor must *theorize* regarding results.

Ultimately, surveying theory and practice must merge as one. They become inseparable if survey measurements are treated properly. Measurement is not merely observing readings on scales and dials or manipulating instrument knobs. Measurement *is* both an art and a science. Without a sound theoretical understanding of measurement, data observed become only so many numbers on a field book page. The initial gathering of data, and certainly the eventual manipulation and reduction of the data can be done carelessly or incorrectly if surveyors misunderstand the explanations of Chapter 2 and those to follow in this chapter.

One intent of this chapter is to provide concepts for understanding and appreciating measurement. In this sense it is theoretical background, each element of which is a further building block toward proper analysis of measurement. A second intent is to illustrate that much of the theory does, in fact, merge with practice.

This chapter focusses primarily on random errors, the theory of systematic errors already having been discussed in Section 2-6.

4-2 SIGNIFICANT FIGURES

The Significance of Significant Figures

A consideration of significant figures can be separated into two areas: (1) significant figures in measurements, (2) significant figures in computations. The first of these relates to the attention the recorder of data gives to reading and recording the values. The second concern relates to the handling of data during computations to avoid round-off errors and extra work with unnecessary digits, and to properly represent final computed results. The two concerns tie together.

In the consideration of *significant figures* (also called *significant digits*), a familiar theme occurs – no measurement is exact. In any measurement, the exact number required to describe the true value would contain an infinite string of digits. During the observation process, the surveyor should always strive to determine just how many digits have meaning and record accordingly. This is important since the number of digits recorded indicate something about the precision of the measurement to the user of the data, and because round-off errors can result in later calculations if insufficient digits have been recorded. The precision of electronic calculators and computers which carry several decimal places will not overcome the improper round-off of the input data. But, recording too many digits, if they have no meaning, adds nothing to final computational precision, misrepresents the precision of the equipment and method, and wastes time and field book space. The determination of the number of figures which are meaningful in a measured quantity, the attention to the number carried in intermediate steps, and the proper representation of final computed results is important, yet these refinements in measuring are given very little attention by some surveyors and engineers. This oversight is inconsistent with good practice. It can result in the precision of a survey being more at the mercy of the recorder of data or computation technician than of the equipment and methods! In summary, it can be said that an appreciation of significant figures is an important part of understanding measurement itself.

Significant Figures in Measured Quantities

Textbooks often state definitions of significant figures such as "the number of digits that have meaning," or "those that are read directly, plus the first doubtful one." Such definitions do not reveal that the "meaning" or "doubtful" part is influenced by the individual observer and by the equipment's precision (least count, other factors). The following is a more complete definition of significant figures in measured quantities:

> *The number of significant figures in any measured quantity is the number that have meaning as estimated from the order of precision with which the quantity was measured, said meaning being limited by the judgment of the person evaluating the observation and the precision.*

For values that are read directly and are not comprised of an accumulation of readings or computed values, the significant figures generally include those read directly and the first doubtful figure. The doubtful figure is generally the one that is estimated rather than read directly. For example, if a circle is divided into minutes of arc and the observer estimates to one-tenth of this least count, the data are significant to the tenths of a minute.

The last phrase in the above definition is important, since it must be realized that each individual could conceivably vary in his or her estimate of precision, and thus, precision and the meaningful digits are always subject to individual judgment and care.

In the process of *making measurements,* the surveyor should observe the following basic rule:

> *The person taking a reading should always record the value to the proper number of significant figures since recording too many implies false precision and too few implies false crudeness.*

In both recording data and interpreting recorded data, the following rules apply:

1. Zeroes used merely to indicate the position of the decimal point are not significant. Example: 0.00618 has 3 significant figures.
2. If zeroes are recorded at the end of a measurement, they are significant.
   Examples: 61.410 has 5 significant figures
   0.0060 has 2 significant figures
3. Zeroes between non-zero digits are significant.
   Examples: 12.1003 has 6 significant figures.
4. In a number ending with one or more zeroes to the left of the decimal, a special indication of the exact number of significant figures must be made.
   Example: $61\bar{5},00$. has 3 significant figures
   $615,00\bar{0}$. has 6 significant figures
   360. could have either 2 or 3 significant figures.

Significant Figures in Computations

In making computations employing measured quantities, a primary concern is to avoid round-off errors which can affect the precision of final results, particularly when the calculations involve several steps. Wasted time in punching extra digits into calculator key boards or onto computer cards should also be of concern. However, one should not waste more time deciding how many digits to use than it would require to punch an extra digit or two into the system. Also of concern is the expression of the final value. Just as too many digits implies false precision and too few implies false crudeness in original measurements, the same is true for calculated results. The final expression of the computed quantity can be made properly only through an analysis of the significant figures in individual computational steps. The following rules for significant figures in computations will aid in these analyses:

1. In adding or subtracting, the number of significant figures in the sum or difference is determined by the fewest decimal places in the numbers added or subtracted. Example:

   14.623
   12.01
   1.0 ← (control—fewest decimal places)
   27.633 → 27.6 having 3 significant figures

2. In multiplication or division, when all numbers are measured values, the number of significant figures in the product or quotient is determined by the

fewest number of significant figures in the values used. Example:

$$14.29 \times 0.051 = 0.73 \text{ (2 significant figures)}$$

3. Conversion factors or constants are not measured quantities, and thus do not determine significant figures. Example:

   $$1342.5 \text{ inches} \div 12 \text{ inches/foot} = 111.87 \text{ feet.}$$

   (The answer in feet has the same number of significant figures as in inches.)
4. To avoid round-off errors when using conversion factors or constants that contain a large or infinite number of digits (or are not counts) use one extra figure in such values. Examples include π and the trigonometric functions. Example involving π:

   $$165.41 \text{ feet} \times 3.14 = 519.39 \text{ (Incorrect)}$$
   $$165.41 \text{ feet} \times 3.14159 = 519.65 \text{ (No round-off error).}$$

   It is noted that using any more than one extra digit in π will not affect the answer in the last place.
5. To avoid round-off errors when using computed values in subsequent calculations, carry one extra figure throughout the intermediate calculations. Example:

   $$131.46 \times 32.68 = 4296.$$

   as a final answer, but use 4296.1 in any further calculations.
6. Round off final answers to the significant figures warranted by the measurements and the rules on computing as cited above.
7. When using an electronic calculator or computer, Rule 5 above (*only*) is unimportant; Rule 4 is unimportant if the constants or functions are generated to more than sufficient precision in the calculator. When calculations are very lengthy, the one additional figure may be insufficient.

Significant Figures and Round-Off Errors

The example on the next page, a sum of nine products, illustrates the important rules above and the manner in which round-off errors can occur. It will be noted that the seventh product has only four significant figures and the remainder have five. The first column contains products which were rounded to only the significant figures dictated by Rule 2, not adding an extra one as in Rule 5. The second column of products applies Rule 5, carrying an extra figure for the products which are the "intermediate" calculations prior to the final calculation of adding the products. The third column carries the products to one more figure beyond that dictated by Rule 5. In the example it can be observed that:

1. If each product is rounded only to the number of figures warranted as though it was a final answer, round-off errors result, the answer being $2{,}575{,}6\bar{2}0$.
2. If one extra place is carried throughout intermediate computations, round-off error is avoided and the answer is $2{,}575{,}\bar{6}00$.

3. There are 6 significant figures in the final answer, not 4 or 5, and not 7.

From this series of computations the following principles can be formulated:

1. Carry one more figure throughout any and all intermediate calculations than would be shown in the final answer to avoid round-off errors.
2. There is ordinarily no need to carry more than one additional figure to avoid round-off errors.
3. The least number of figures in one number used in multiplication bears little relationship to the significant digits in a summation of several products since the rule for addition applies for the summation.

These rules and principles have many applications in surveying, some of which are beyond the scope of this text. They are especially applicable in analyzing measurements, the primary subject matter for the remainder of this book.

	least no. only	least no. +1	least no. +2
431.21 × 400.00	$1724\bar{8}0$	172484.	172484.0
398.38 × 200.00	$7967\bar{6}$	79676.0	79676.00
635.42 × 300.00	$1906\bar{3}0$	190626.	190626.0
981.33 × 100.00	$9813\bar{3}$	98133.0	98133.00
948.39 × 500.00	$4742\bar{0}0$	474195.	474195.0
698.21 × 700.00	$4887\bar{5}0$	488747.	488747.0
66.92 × 800.00	$535\bar{4}0$	53536.	53536.0
139.44 × 900.00	$1255\bar{0}0$	125496.	125496.0
1785.42 × 500.00	$8927\bar{1}0$	892710.	892710.0
Σ =	2575619	2575603.	2575603.0

Significant Figures as a Precision Index

Lacking statistical analyses or other means, a level of precision can be attached to a measurement by examining the digits recorded. The precision is generally considered to be ± ½ the last place. For example, 42.81 is understood to be 42.81 ± 0.005. From this, it can also be reasoned that it would be improper to write the number as 42.81 if the error was closer to ± 0.05. In this case, 42.8 would be appropriate instead.

Reversing the above logic, one can discover how many significant figures are contained in a measured quantity if the precision is stated. For example, if the above measurement had an uncertainty of ± 0.5, one would know that the recorded measurement should end in the units column, the value being 43. (two significant figures). This logic must be less strictly applied in the usual case when the uncertainty is not exactly ½ of a decimal place. For example, consider quantities having uncertainties of ± 0.07 or ± 0.3. Should they be recorded to tenths or units? Common sense would say that precision should not be sacrificed and the quantities should be recorded in tenths: 42.8 ± 0.07 and 42.8 ± 0.3, etc. We might then say that this number is closer to having three significant figures than two.

4-3 ESTIMATES OF UNCERTAINTY

From the above, a meaning of *precision index* should be deduced. Any plus-minus estimate is a precision index. The term is often considered synonymous with *random error* which is also the same as *uncertainty.* All three terms, when used in the context of error estimates mean *the positive and negative range of values expected for a recorded or computed value.* A statement of uncertainty is a statement of the precision of the measured or computed value. It is an assertion that the value shown is not exact and is an attempt to indicate the inexactness. Such a statement must always be accompanied by plus or minus (±).

If no statement of uncertainty is given with a measurement the user of the data is forced to rely on the convention that the precision is ± ½ the last decimal place. It also forces the user to have faith that the person who recorded the data or made the computations understood all of the rules on significant figures.

If the significant figures convention results in an inaccurate indication of uncertainty, it can be overridden by including other statements of uncertainty. Such statements can be based on (1) judgment, or (2) statistical analysis.

Judgment, as used here means the professional level of judgment which is well considered, based on experience, and exercised only after understanding the theoretical concepts. It is not uneducated guessing or uninformed opinion or whim. It is the type of judgment or opinion any professional person must make when exact answers are unknown. In surveying, it is common sense based on knowledge of measuring methods, field conditions, equipment capabilites, computational methods, and geometric characteristics of a particular survey. A surveyor may be able, from experience and knowledge of a particular survey, to estimate a range of uncertainty of a measured quantity. For instance, the surveyor may have measured a mile by taping and realizing that the field methods have yielded accuracies of about 1 part in 10,000 feet in the past, make a statement of ± 0.5 feet for an uncertainty. If the actual field reading, with subsequent corrections for systematic errors yielded a value of 5281.47 feet, a complete statement regarding this measurement would be 5281.5 ± 0.5 feet.It is noted that the hundredths in the reading are meaningless but there is nothing wrong with giving the value to tenths *if* the uncertainty estimate is added. If not, the tenths would indicate false precision (± 0.05) and should not be shown.

Much of the remainder of this chapter will be devoted to explaining the second approach to estimating uncertainty in measurements — that of statistical analysis.

4-4 ANALYSIS OF REPEATED OBSERVATIONS

Random Error Theory

The primary cause of random errors is the inability of the observer to make exact measurements. Although perhaps attempting to be exact, the observer will read, point, or otherwise respond somewhat to one side of the true value, sometimes to the other side, the errors varying in magnitude, and rarely if ever being zero. The algebraic sign of random errors is, by nature, plus or minus. Due to the accidental nature of random errors, small errors are more likely to occur than large ones. This likelihood fits into a definite theoretical pattern with a large sample of repeated measurements of a quantity, provided the measurements are made with the same care each time. In order to clearly observe this pattern from a plot of a finite and manageable number of repetitions, it is important that the instrument or method be more refined than the ability of the observer to read the scale. For example, something different would be observed with each measurement of an angle if a one-second theodolite was used, but it is likely that the same reading would be observed each time if a one-minute theodolite was used. As one author stated this phenomenon — "more precise methods tend to magnify discrepancies or stated differently, crude methods tend to hide discrepancies." (Ref. 1)

With a sufficiently large sample of readings of a single quantity, assuming no bias or changes in methods of observation, the distribution of errors would fit the theoretical characeristics of random errors (i.e., small errors occurring more frequently than large ones, as much likelihood of errors being plus as minus, etc.). Graphs and curves which reveal this graphically will be explained.

Averages

Three terms that have a close, but different meaning are the *mean*, the *median*, and the *mode*. All three are ways to express "average" in a general sense and are sometimes called measures of central tendency. The *arithmetic mean* is the sum of the individual observations divided by the number of repetitions. The *median* is simply the middle observation when the data are arranged in ascending or descending order, or the mean of the two middle observations if n is an even number and the two are different. The number of observations which are larger than the median is equal to the number that are smaller. The *mode* is the value which occurs most frequently.

If a sample set is truly random, the mean, the median, and the mode should all be the same value. They may be slightly different if there is a skew to the histogram or probability curve, or if there are a few *minor modes* (humps in the probability curve other than at the central mode). Studying these differences is another means of evaluating the randomness in a set of repeated observations. Explanation of histograms, probability curves and how they relate to these terms will follow.

Histograms

A histogram is a bar graph of repeated observations of a quantity, the readings being plotted on the abscissa and the number of occurrences (frequency, f) of each value being plotted on the ordinate. The abscissa interval (i.e., whether divisions are tenths, units, etc.) is called the *class interval*. Proper choice of this interval facilitates interpretation of data.

An example of a theodolite reading error test will be used to illustrate the construction and usefulness of a histogram. Table 4.1 shows repeated readings taken using an optical theodolite with the coincidence type reading system shown in Fig. 4.1 (a). A coincidence knob is used to align scale marks, after which the reading is observed on an adjacent scale. The divisions on this second scale are to seconds and estimation is to 0.1 seconds. The reading systems of theodolites are more thoroughly discussed in Section 6-3.

No angles were measured nor any pointings made in this test, it being made merely to study the reading capabilities of the instrument-observer combination. The optical microscope was focussed and one reading observed. Next, the coincidence knob was used to again misalign the scale marks, then realign them and record the second observation. After 25 observations, the set of data in Table 4.1 was completed. Table 4.2 shows the grouping of the readings in three ways: (1) According to frequency of occurrence in the intervals as actually estimated (to nearest 0.1″), (2) grouped into intervals of 0.5″, and (3) grouped into intervals of 1″. Figure 4.2 shows the histograms of these three groupings.

If the data truly follow the theory of random errors, there should be a tendency toward grouping around the mean or median reading. This is scarcely seen in Fig. 4.2 (a), but becomes more apparent in Figs. 4.2 (b) and (c). Remember what was said earlier about the importance of having an instrument which is more refined than the ability of the observer to observe, and that the more precise the method, the more the discrepancies are magnified. The "method" of reading to tenths certainly magnified discrepancies as they become spread across the range. But, had readings been estimated to 0.5 seconds or 1.0 seconds, these discrepancies would not have been so apparent. Also, it could be imagined that if the instrument had been constructed with minutes instead of seconds or even divisions into 10 seconds instead of 1 second, the histogram might well have been only one value with a frequency of n, the total number of observations. That is, any discrepancies would have been hidden completely because of the crudeness of method. Another observation from these data, which should become more apparent as the subject is developed, is that there is a point reached in the design of certain types of instrument scales where further refinement in the scale design is wasted since observer ability, combined with other error sources, limits the precision more than scale refinement alone.

Let us next examine the distribution of two other sets of theodolite readings. Table 4.3 shows distribution and grouping for 25 repetitions of a circle reading using an instrument similar to that used for Set 1. The difference in the reading systems can be seen by comparing Fig. 4.1 (a) with Fig. 4.1 (b). The digital readout is a newer scale system that first appeared on the Wild T-2 around 1972. Table 4.4 shows distribution and grouping for 25 repetitions of a circle reading using a theodolite having a reading system similar to the T-2. The procedures for collecting data for Sets 2 and 3 were identical to those used for Set 1, all instruments having a coincidence knob for aligning scale markings. The histograms for Sets 2 and 3 are shown in Figs. 4.3 and 4.4. Plots with class intervals of 0.1 seconds were omitted. They would have appeared scattered similar to Set 1 as can easily be seen upon examination of the distribution in the tables. The histograms of these two sets, when grouped into class intervals of 0.5 and 1.0 seconds begin to reveal the clustering about the values approximating the mean or median and also form into the familiar shape of a random set, with a peak near the mean, smaller ordinate values as reading values become farther from the mean, and a general symmetry as the data begin to behave according to the equal probability of errors being positive as negative.

Much more will be discussed and compared using these histograms as examples. First however, a mathematical curve which has been developed by statisticians to represent random data will be explained.

The Normal Distribution Curve

This curve, commonly known as the Bell curve or Gauss curve (and, also normal probability curve, or simply "probability curve"), has a mathematical equation which equates the frequency of occurrence with variables which are a function of the precision of the method. The equation of the curve is rarely used in surveying measurement analysis, but various properties to be used will be explained. It can be proved mathematically that if an infinitely large random sample was plotted in histogram form, and the center point of each ordinate connected by a smooth curve, the result would be the Gauss curve. Therefore, for a fairly large sample, a histogram does approximate the general shape of the Gauss curve. To illustrate this, Fig. 4.5 shows a curve superimposed on a histogram of a hypothetical random sample.

It is to be noted that the units on the abscissa of a histogram are the various readings themselves, scaled to a selected class interval. Henceforth, however, the abscissa for the probability curve will have units of errors (actually residuals), measured with respect to the line of symmetry (representing the mean) as shown in Fig. 4.6. The ordinates remain as numbers of occurrences (frequency) of errors.

It is important at this point to distinguish between *error* and *residual*. As has been stated, error is the difference between a particular reading and the true value. But, the true value in a set of repeated readings of a quantity is unknown; therefore, the errors are also unknown. The *arithmetic mean* is often called the "best value", since it is the best available representation of the true value. A *residual, v,* is the difference between a particular reading in a set of repeated measurements of a quantity and the arithmetic mean.

$$v = \text{Reading} - \text{Mean} \tag{4.1}$$

While we will often speak of distribution of errors about the mean when discussing the probability curve, in actuality the abscissa of the curve has units of residuals rather than errors, measured as either plus or minus from the line of zero residuals (the mean or best value for the set).

With the similarity and close relationship between the histogram and the probability curve fresh in mind, it is well to note at this point that the entire set of repeated readings is fully represented under the graph or curve. That is, there is no measured value or residual lying outside of the curve and therefore the sum of the values under the curve or beneath the bars represent 100% of the occurrences. Since, theoretically the probability curve approaches the x axis at plus and minus infinity, the probability of a residual occurring within these limits is 100%. The area under the curve between plus and minus equal residuals of very large size is nearly equal to, but always less than unity or 100%. Thus, it becomes obvious by intuition that the area under the curve between any selected plus and minus, equal sized residual values represents the probability that any value, selected at random

TABLE 4.1

THEODOLITE MICROMETER READINGS

SET 1

Wild T-2 (Coincidence Circle)

Observation	Reading, Seconds
1	31.8
2	32.3
3	31.9
4	30.2
5	31.4
6	32.7
7	31.1
8	31.0
9	31.2
10	30.3
11	30.4
12	30.9
13	31.4
14	31.0
15	31.4
16	30.8
17	31.8
18	31.0
19	30.6
20	30.4
21	31.4
22	30.2
23	31.0
24	31.5
25	31.7

Note: All readings are in seconds.

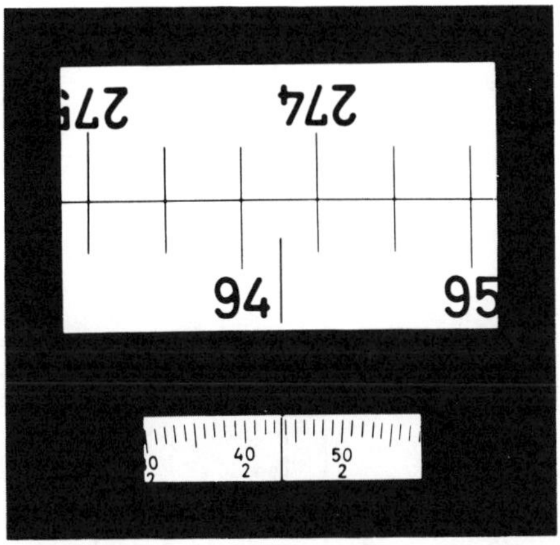

360° Readings: of vertical or horizontal circle:
94° 12′ 44″

(a)

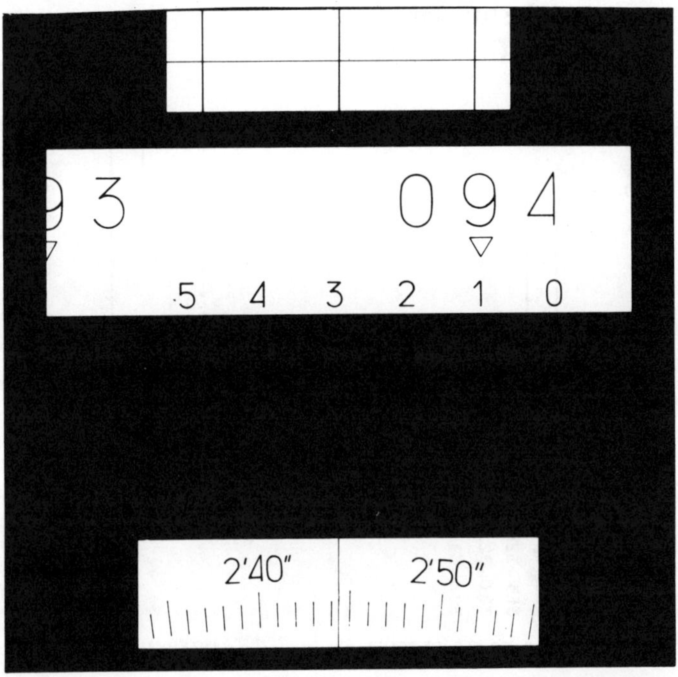

Vertical circle reading (360°) 94° 12′ 44″.

(b)

Figure 4.1 Direction Theodolite Reading Systems.

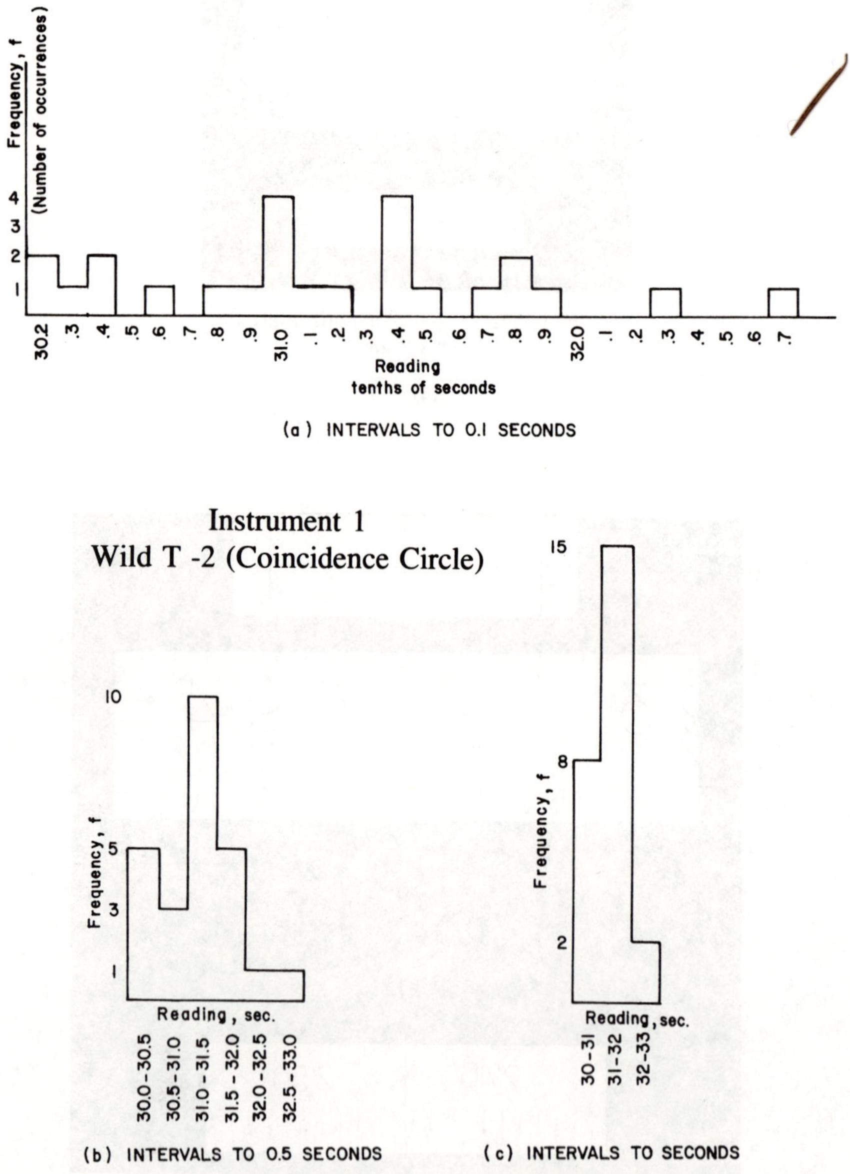

Figure 4.2 Histograms of Theodolite Readings - Set 1

TABLE 4.2

DISTRIBUTION OF THEODOLITE DATA

SET 1

Wild T-2 (Coincidence Circle)

Actual Readings	f
30.2	2
30.3	1
30.4	2
30.5	
30.6	1
30.7	
30.8	1
30.9	1
31.0	4
31.1	1
31.2	1
31.3	
31.4	4
31.5	1
31.6	
31.7	1
31.8	2
31.9	1
32.0	
32.1	
32.2	
32.3	1
32.4	
32.5	
32.6	
32.7	1
	n =25

Grouped into Intervals of 0.5″

Grouped Readings	f
30.0 - 30.5	5
30.5 - 31.0	3
31.0 - 31.5	10
31.5 - 32.0	5
32.0 - 32.5	1
32.5 - 33.0	1

Grouped into Intervals of 1″

Grouped Readings	f
30 - 31	8
31 - 32	15
32 - 33	2

Mean = 31.2
Median = 31.1
Modes = 31.0 and 31.4
Range = 2.5
Modes:
(in grouped intervals) = 31.0 - 31.5 and 31 - 32 ranges

Note: All readings are in seconds.

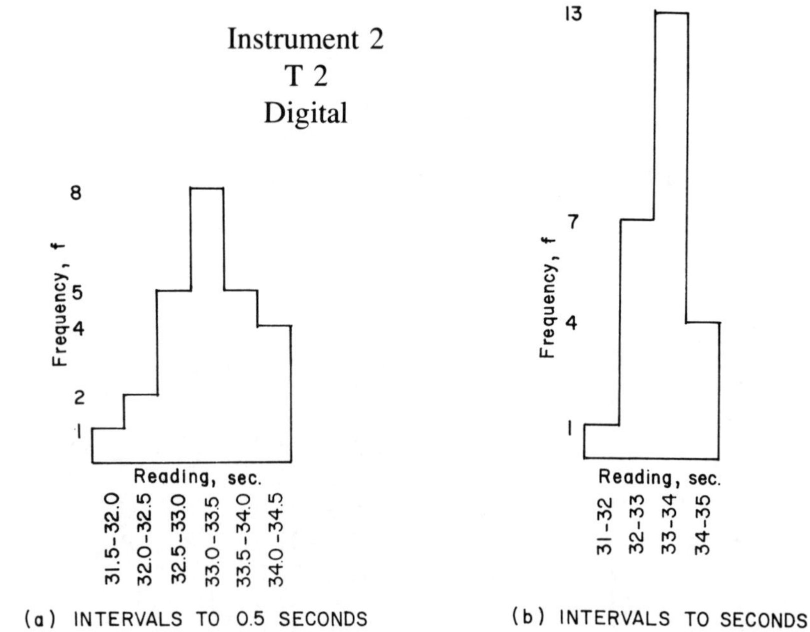

Figure 4.3 Histograms of Theodolite Readings - Set 2

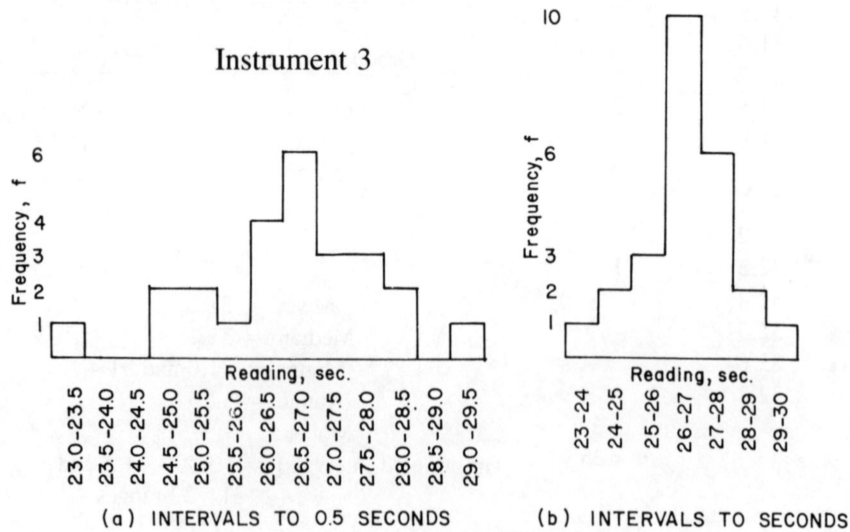

Figure 4.4 Histograms of Theodolite Readings - Set 3

TABLE 4.3

DISTRIBUTION OF THEODOLITE DATA

SET 2

Wild T - 2 (Digital)

Actual Readings	f
31.8	1
31.9	
32.0	
32.1	
32.2	
32.3	
32.4	2
32.5	2
32.6	1
32.7	
32.8	1
32.9	1
33.0	2
33.1	2
33.2	2
33.3	1
33.4	1
33.5	2
33.6	1
33.7	
33.8	2
33.9	
34.0	3
34.1	1
	$n = 25$

Grouped into Intervals of 0.5″

Grouped Readings	f
31.5 - 32.0	1
32.0 - 32.5	2
32.5 - 33.0	5
33.0 - 33.5	8
33.5 - 34.0	5
34.0 - 34.5	4

Grouped into Intervals of 1″

Grouped Readings	f
31 - 32	1
32 - 33	7
33 - 34	13
34 - 35	4

Mean = 33.2
Median = 33.2
Mode = 34.0
Range = 2.3
Modes:
(in grouped intervals) = 33.0 - 33.5 and
33 - 34 ranges

Note: All readings are in seconds.

TABLE 4.4

DISTRIBUTION OF THEODOLITE DATA

SET 3

Actual Reading	f	Actual Reading	f	Acutal Reading	f
23.4	1	25.3	1	27.2	1
23.5		25.4	1	27.3	
23.6		25.5		27.4	1
23.7		25.6		27.5	1
23.8		25.7		27.6	1
23.9		25.8		27.7	
24.0		25.9	1	27.8	
24.1		26.0		27.9	1
24.2		26.1		28.0	
24.3		26.2	2	28.1	
24.4		26.3	1	28.2	
24.5		26.4	1	28.3	2
24.6	1	26.5		28.4	
24.7		26.6	2	28.5	
24.8		26.7	2	28.6	
24.9	1	26.8	2	28.7	
25.0		26.9		28.8	
25.1		27.0	1	28.9	
25.2		27.1		29.0	
				29.1	1

Grouped into Intervals of 0.5″

Grouped Readings	f
23.0 - 23.5	1
23.5 - 24.0	0
24.0 - 24.5	0
24.5 - 25.0	2
25.0 - 25.5	2
25.5 - 26.0	1
26.0 - 26.5	4
26.5 - 27.0	6
27.0 - 27.5	3
27.5 - 28.0	3
28.0 - 28.5	2
28.5 - 29.0	0
29.0 - 29.5	1

Grouped into Intervals of 1″

Grouped Readings	f
23 - 24	1
24 - 25	2
25 - 26	3
26 - 27	10
27 - 28	6
28 - 29	2
29 - 30	1

Mean = 26.6
Median = 26.7
Mode = Several
Range = 5.7
Modes:
(in grouped intervals) = 26.5 - 27.0 and 26 - 27 ranges

Note: All readings are in seconds.

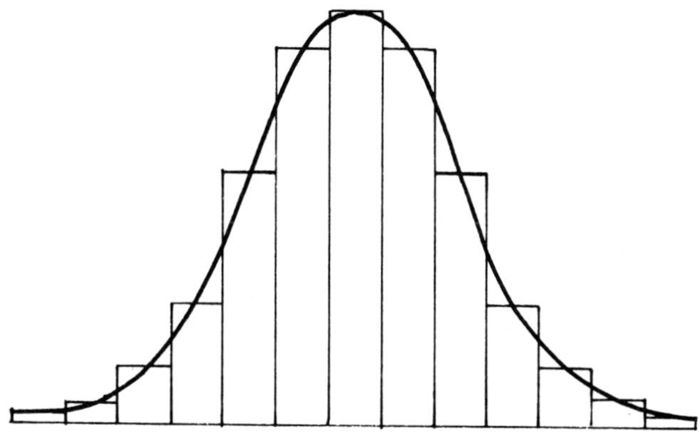

Figure 4.5 Histogram and Probability Curve

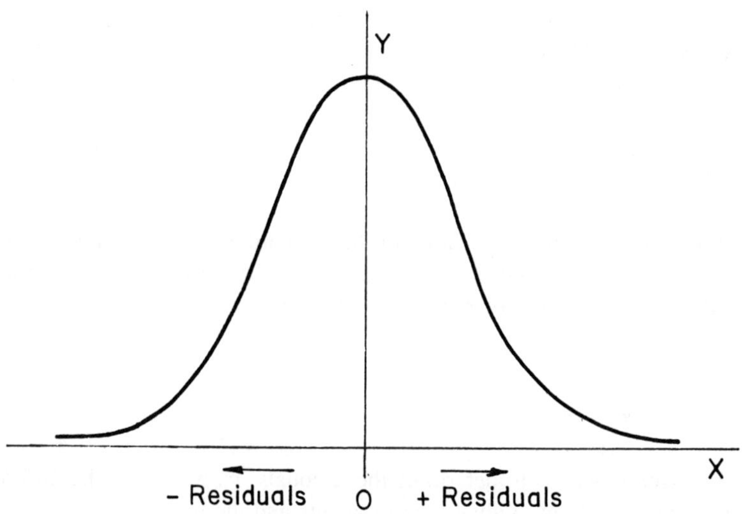

Figure 4.6 Normal Probability Curve

from the set, will lie within those limits selected. This can easily be observed from an examination of data plotted on a histogram. Here, actual count of the readings which fall within selected plus and minus differences from the mean, divided by the total number of observations, will be a ratio equal to the probability of any randomly selected value falling within that range. This property of the curve provides a tool which can be used to analyze measurement data and predict important qualities about the measurements. Examples illustrating this concept are included in the following sections.

The Standard Deviation

The area under a probability curve between the plus and minus points of inflexion represents 68.3% of the total area under the curve (see Fig. 4.7). These two residual values are known as the *standard deviation*, σ, which is calculated from the equation

$$\sigma = \sqrt{\frac{\Sigma \nu_i^2}{n-1}} \tag{4.2}$$

where $\Sigma \nu_i^2$ is the sum of the squares of the residuals (one residual for each observation) and n is the total number of observations. The derivation of this formula is beyond the scope of this text. The fact that the σ values occur at the points of inflexion is not particularly important in analyzing measurements and the fact that the probability is 68.3% is primarily of academic interest. However, the meaning of percent probability, regardless of the value, is important for the student to conceptualize. Theoretically, there is a 68.3% chance that any *single* value, selected at random from the entire set of observations, will fall within the plus and minus range represented by the standard deviation. For this reason, σ is often called the *standard deviation of a single value*. Some statistics texts use n rather than $n-1$ in the denominator. The difference in results in σ is negligible when n is large.

The first step in the calculation of the standard deviation is to determine the arithmetic mean, $\bar{x}$, *which is equal to the sum of the individual measurements, divided by* n, the number of repetitions of the measurements.

$$\bar{x} = \frac{\Sigma x_i}{n} \tag{4.3}$$

The second step is to determine the residuals, ν_i, which are the differences between each of the individual measurements and the mean.

$$\nu_i = x_i - \bar{x} \tag{4.4}$$

The third and fourth steps are to square the residuals and to sum them. The last step is the calculation of σ using Eq. 4.2.

Data Set 1 of the theodolite readings will be used to demonstrate the computation of σ. The first two columns in the set of calculations of Table 4.5 are identical to

TABLE 4.5

CALCULATION OF STANDARD DEVIATION

SET 1

Wild T - 2 (Coincidence Circle)

Observation	Reading	ν	ν^2
1	31.8	+0.6	0.36
2	32.3	+1.1	1.21
3	31.9	+0.7	0.49
4	30.2	−1.0	1.00
5	31.4	+0.2	0.04
6	32.7	+1.5	2.25
7	31.1	−0.1	0.01
8	31.0	−0.2	0.04
9	31.2	0.0	0
10	30.3	−0.9	0.81
11	30.4	−0.8	0.64
12	30.9	−0.3	0.09
13	31.4	+0.2	0.04
14	31.0	−0.2	0.04
15	31.4	+0.2	0.04
16	30.8	−0.4	0.16
17	31.8	+0.6	0.36
18	31.0	−0.2	0.04
19	30.6	−0.6	0.36
20	30.4	−0.8	0.64
21	31.4	+0.2	0.04
22	30.2	−1.0	1.00
23	31.0	−0.2	0.04
24	31.5	+0.3	0.09
25	31.7	+0.5	0.25
Σx_i = 779.4		$\Sigma \nu^2$ = 10.04	

$$\bar{x} = \frac{779.4}{25} = 31.2$$

$$\sigma = \sqrt{\frac{\Sigma \nu^2}{n-1}} = \sqrt{\frac{10.04}{24}} = \sqrt{0.418} = \pm\ 0.65$$

Note: All readings are in seconds.

Table 4.1. The student should check the application of the five-step process just explained for calculation of σ. Most hand-held scientific calculators are preprogrammed to compute $\bar{x}$ and σ. However, the student should gain an understanding of the theory by first using the five-step process just explained before using more automated methods.

From the theory, it can now be stated that any reading selected at random from Set 1 has a 68.3% chance of falling in the range 31.2 $\pm$ 0.65″. Or, approximately 68% of the readings (in this case 17) should be in the range 30.55 to 31.85. That would include all readings 30.6 through 31.8. A count reveals that there are actually 17 readings in that range. Thus, the data passes one elementary test of being close to a true random sample. Had the count been very much different from that expected, bias or other factors causing the sample to be non-random would be suspected. For a graphical observation of the readings falling in the σ range, the student should return to Fig. 4.2 (b). Note that there were 10 readings in the range 31.0 to 31.5, 3 more in the 30.5 to 31.0 range, and 5 more in the 31.5 to 32.0 range. These 18 represent a range only slightly wider than $\pm\ \sigma$.

From a simple test of σ from repeated measurements of a quantity, the uncertainty of any future single measurement of the same quantity or another quantity using the same methods, can be reasonably estimated. This statement must be made and accepted with caution, however. It is only true if (1) the conditions (including equipment, procedure, and environmental conditions) existing for the test sample are the same for the later single observation, and (2) there are no systematic errors or mistakes (only random errors), in any of the observations. If the same care and equipment is used, it is a reasonable assumption that the later single observation has an uncertainty equal to that yielded by the test. If the second condition is true, the arithmetic mean for an infinite size sample should be the true value and thus a statement of σ is a statement of uncertainty of a value with respect to the true value. This will be more completely explained in the next section, the main point here being that σ is a precision index which is a reliable method for stating uncertainty in measurements. Discussions of σ will continue throughout much of the remainder of the text. The reader may need to return to this section several times as the full significance of this precision index is developed.

Standard Error of the Mean

The *standard error of the mean,* $\sigma_{\bar{x}}$, is calculated as follows:

$$\sigma_{\bar{x}} = \pm \frac{\sigma}{\sqrt{n}} \tag{4.5}$$

where σ and n are as defined previously. From this equation, it should be apparent that the mean approaches the true value as n increases. Indeed, if $n = \infty$, $\sigma_{\bar{x}} = 0$. The standard error of the mean is an uncertainty statement regarding the *average* (mean) of the set and *not* of a randomly selected *single* value as is σ. Furthermore, the uncertainty is with respect to the *true* value, not the mean itself. The statement that the mean approaches the true value is valid only if errors are purely random in nature (i.e., systematic errors have been corrected and there are no mistakes). If a constant calibration error existed throughout a series, the mean would differ from the true value by the amount of that error, for example, and no amount of repetitions could cause the mean to approach the true value.

In that σ has a 68.3% probability attached to it, so does $\sigma_{\bar{x}}$. That is, the mean has a 68.3% probability of being within $\pm\sigma_{\bar{x}}$ of the unknown true value.

It will be seen that Eq. 4.5 is a powerful tool in analyzing measurements and particularly in the design of measuring systems. It provides a ready means to either (1) estimate the theoretical number of measurements required in order to achieve a specified uncertainty in the mean of a set of measurements using a given instrument, or (2) select an instrument which will require a minimum, practical number of repetitions for achieving a specified uncertainty. Examples of this application of the equation will appear in chapters to follow.

Level of Certainty

The σ or $\sigma_{\bar{x}}$, with their inherent 68.3% probability, are perhaps not the most useful precision indexes. No surveyor would risk the liability of measuring with only approximately ⅔ confidence in uncertainty statements. If expensive structures or high land values are involved, a professional surveyor might rest easier at 90% or even 99%. Percent probability will hereafter be called *level of certainty.* It represents a level or degree of confidence regarding an error or uncertainty statement.

Any desired level of certainty can be calculated from σ. Common levels used are as shown in Table 4.6. Another often used term, not appearing in the table is "maximum" error, given a symbol of E_{max}. It is usually defined as 3.29 σ and has a 99.9% certainty or only a probability of 1 in 1,000 that a value would be outside of the range. Some authorities regard the slightly smaller 95% error as the "maximum error." Obviously, all such attempts to define a maximum error in finite terms are impossible since it actually occurs at $\pm\infty$, the limits represented by 100% of the area under the curve.

Some of the levels of certainty are graphically illustrated in Fig. 4.7. Note that the percent area under the curve between specific uncertainty values is the level of certainty of that error. For example, the cross-hatched portion would be 68.3% of the total area. Just as with σ, the number of actual observations between error limits with a specific level of certainty can be counted to develop a feeling for the concept. For Data Set 1, $E_{90} = \pm 1.06$. There are 23 observations falling in the range 31.2 ± 1.06. This is 92% of 25, as close to the theoretical 90% as is possible with this value of n.

The practitioner who has worked for many years with complete confidence in his measurements should pause at this point to ponder the truths implicit within the concept of *level of certainty.* One can be anything *less than* 100% certain of any precision level of measurements. How certain, as between practical maximums of perhaps 95 to 99% is an individual professional decision. In some instances, 90% or even as low as 50% certainty might be appropriate. As will be seen in later chapters, to increase the level of certainty for any particular uncertainty value requires more effort or expense (i.e., more repetitions or more precise instruments). Applications of these ideas to practice will be discussed in Chapter 9.

The error in the mean for a particular level of certainty can be calculated using the constants in Table 4.6. The 90% error in the mean of a particular set of measurements, for example, would be calculated from the 90% error in a single measurement as follows:

$$E_{90m} = \frac{E_{90}}{\sqrt{n}}$$

Any other error in the mean for a particular level of certainty would be calculated the same way, substituting the respective error in a single measurement in the right side of the equation. The error in the mean will have the same level of certainty as the error in the single measurement.

Table 4.6 adds additional useful tools for analyzing survey measurements.

Size of Sample

After examining the shape of the bar graphs of Figs. 4.2 – 4.4, the reader may have been disturbed at their lack of symmetry and bell shape and wondered how it is possible to put much reliance on precision indexes based on σ or other theoretical indexes. First of all, it must be reemphasized that all such estimates are just that—estimates. Even an uncertainty has an uncertainty! The degree of certainty in the estimate is improved when the histograms appear symmetrical and bell shaped. There is more confidence in the randomness of the sample set when this occurs. Obviously, 25 repetitions were insufficient to build a very symmetrical histogram, even when class intervals were increased. There is a *skew,* for example, in the plots in Fig. 4.2 (b) and (c), there being a few more readings in the lower ranges than in the higher.

So, how many observations are needed to build a symmetrical histogram? The number will vary with the class interval selected. Even using 0.1″, it should be realized that several hundred or thousand repetitions would have filled in the gaps and built a symmetrical histogram for Data Set 1. The reason there is skew to the histograms could be bias, but it could also be simply a lack of sufficient number of observations to counteract the chance happening of a few more lower readings. The entire theories of probability and compensation do not dictate that there must be symmetry at all stages in the sampling operation. But, laws of probability do say, that, with a sufficiently high number of repetitions, patterns will approach their theoretically correct shape.

The choice of how many repetitions to make for a sample test of a measuring system relates to how many are needed in order for the observer to gain an individual level of confidence that the errors are primarily random in nature. The best way to do this is to plot histograms of the data, as in Figs. 4.2 – 4.4. When to stop repeating observations is a matter of professional judgment, using this graphical examination as a basis. A feeling for a practical number of observations to make in a test sample can also be gained by testing the value of σ at various stages of increasing n. If σ fluctuates, n is too small. When σ starts to stabilize, n is probably sufficiently large. Practical problems of cost, time, and human fatigue often limit the value of n. Generally, n should be at least 25, and more if possible.

TABLE 4.6

LEVELS OF CERTAINTY

Name of Error	Symbol	Value	% Certainty
Probable	E_{50}	0.6745σ	50
Standard deviation	σ	1σ	68.3
90% Error	E_{90}	$1.6449\ \sigma$	90
Two-Sigma or 95% Error	E_{95}	2σ	95
99% Error	E_{99}	$2.5\ \sigma$	99
Three-Sigma	$E_{99.7}$	3σ	99.7

(Note: Acutally, the 95% error is closer to 1.96σ, but 2σ is often accepted as a convenient conversion. The conversions for the 50% and 90% errors are normally not needed to the significant figures given in the table.)

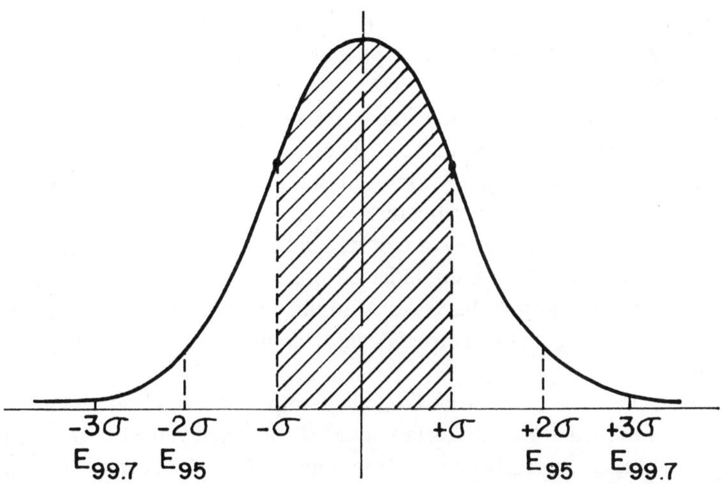

Figure 4.7 Theoretical Precision Indexes

Choice of Class Interval

Comparisons of Data Sets 1, 2 and 3 can be used to illustrate a point about how to select a practical class interval for the histograms. The σ for Set 1 has been calculated. The σ values for Sets 2 and 3 can be calculated in a similar manner. The values for all three are as follows:

$$\sigma_1 = \pm 0.65'' \qquad \sigma_2 = \pm 0.60'' \qquad \sigma_3 = \pm 1.26''$$

Now, examine the histograms for each set. Note that the class intervals of 0.1″ were too spread out to be useful, but when class intervals of 0.5″ or 1.0″ were used, the graphs started to take shape. Note, in particular, the plots for Data Sets 2 and 3. Notice how, even at an interval of 0.5 seconds, Data Set 3 is still rather "roller coaster" in shape but then it achieves a certain compactness and symmetry at a class interval of 1.0″. Notice next the similarity between the shape of Data Set 3 at 1.0 seconds interval and Data Set 2 at 0.5 seconds interval. Note, also that these respective intervals are approximately the σ values (actually each is slightly smaller).

From the above observations, a rule of thumb follows that the class interval selected for a sample of limited size (say, 20–50 repetitions) should be approximately the σ value. Thus, a preliminary σ should be calculated before plotting a histogram and before making final decisions as to whether the sample is large enough as discussed in the previous section.

Analysis of Relative Precision

Precision has been defined as relative agreement among repeated measurements of the same quantity. Better precision occurs when values agree more closely among themselves. It should be obvious that closer agreement results in a lower σ since the sum of the squares of the residuals would be less. Quantitatively, the best way to compare relative precision between sets of measurements is to compute the σ values. The method or instrument resulting in the lower σ would be the more precise. Comparing Data Sets 1, 2, and 3, it could be concluded that the instrument used for Set 2 was slightly (perhaps insignificantly) more precise than the instrument used in Set 1. The scale system of the instrument of Set 2 is presumably an improvement, so this improved σ value is to be expected. Both Sets 1 and 2 have lower σ values than Set 3. So, from this test alone, one might conclude that the reading system employed in the instrument of Set 3 is the least precise. The reader should not erroneously conclude, however, that the particular instrument itself is inferior or even that its reading system is inferior. All three samples are of limited size, taken under different natural conditions of lighting, and by an operator who may have varied in his own personal care. Also, many other instrumental and personal errors occur in angle measurement. Only one has been tested here.

Relative precision can also be compared graphically by observing the shape of histograms plotted at the same class interval and equal value of n for each set being compared. It can easily be seen from Figs. 4.2 – 4.4 that Data Sets 1 and 2 "stack up" higher and tighter at any class interval than Set 3, so the relative precision can be seen even without calculation of σ. The histograms of Data Sets 1 and 2 are very close in structure, so it would be difficult to decide which is more precise by studying their histograms. From these comparisons, it should be obvious that a more precise set of measurements would yield a histogram with a higher peak and narrower base, assuming the same class interval and value for n.

From the above, it follows that normal probability curves with higher peaks represent higher precision than low, flat ones, as shown in Fig. 4.8 where Set A is more precise than Set B.

It should be noted that accuracy, or nearness to the true value, has not been discussed above. If systematic errors occurred in measurements, precision without accuracy could result. Set A in Fig. 4.8, although more precise, is less accurate than Set B since its mean is farther from the true value. Had systematic errors been

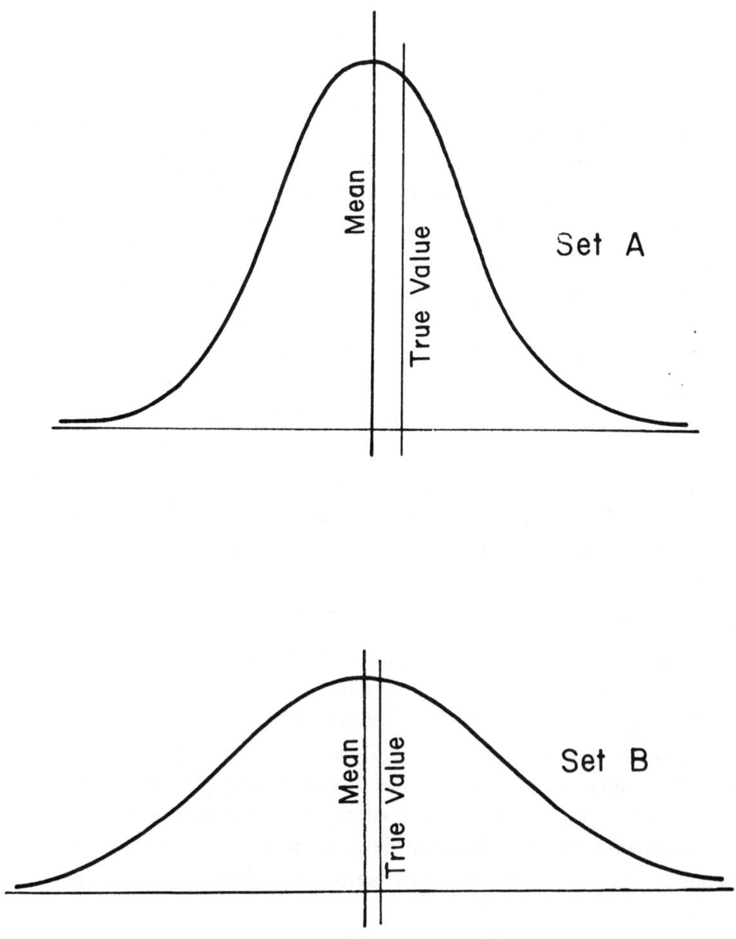

Figure 4.8 Probability Curves as Precision Indicators

corrected in either case, the same patterns would presumably have resulted, but centered around a mean coinciding with the true value. After corrections, all readings would have been higher in either set.

Other Indicators of Precision

By this time the student should be developing a feeling for precision and how to test and compare it by repeated measurements. The experienced practitioner often checks precision by merely observing how repeated measurements agree with each other in the field book and may reject values or entire sets if unsatisfied. This is an intuitive exercise with some, but whether conscious or unconscious of the theoretical basis, the surveyor is using one or both of the following two indicators of precision: the *range* and the *scatter*. These terms were introduced in Section 2.3. The *range* is simply the difference between the highest and lowest values. For example, as shown in Tables 4.2-4.4, the range of Data Sets 1, 2, and 3 were 2.5″, 2.3″, and 5.7″. The range (also called "spread") reveals something of precision because it indicates that errors approaching the magnitude of the range have a reasonable probability of occurring under normal conditions (unless the values at either or both extremes are mistakes or "wild" values). A wider range would result in a higher σ and probably also more scatter. However, range itself does not reveal actual dispersal of readings between the extreme values. The remainder of the values could conceivably be tightly clustered. Therefore, both range and scatter should be used when examining data. *Scatter* is the nature in which the readings distribute themselves between the extreme values. The set as a whole must be studied in order to observe scatter. A study of Tables 4.2 – 4.4 show the much wider dispersal of Data Set 3, there being more gaps in the data and relatively few readings even repeating themselves twice. The respective numerical ranges of the three data sets are consistent with former conclusions as to their relative precision. The scatter, as seen in the tables, is also consistent with these conclusions. From this discussion, it should be apparent that a general idea about the precision of a set of observations is possible without either plotting histograms or calculating σ, although the latter are considered to be more thorough and conclusive.

Rejection of Observations.

The observations rightfully belonging to a true "family" of random readings are those that contain no mistakes or systematic errors. The complete family includes the large random errors that "rarely occur". But, how does one know whether the readings at the extremes should be part of the family or thrown out as "misfits"? The only justification for rejecting such observations is that a mistake, fault in the instrument, or other temporary condition caused the value to be classified as a *wild value*. Obvious mistakes, occasional carelessness, a change in environmental factors, temporary distraction, or other unusual influences can cause wild values.

A simple examination of the data may reveal wild values. If a value is completely alone, separated by several gaps in the distribution table, it might be readily classified as a wild value. Such might be the case for the reading of 23.4 in Data Set 3. It could also be the case for the 29.1 reading in that set or the 31.8 in Set

2 or the 32.7 in Set 1. But, as can be seen, one could easily start trimming any data set of its extreme readings until it satisfies his whims for neatness or symmetry. Although occasionally a clear mistake can be detected and deleted this way, readings varying because of carelessness or simply because they seem "out of place" in a small sample, should not be rejected based on arbitrary judgments. Although still somewhat arbitrary, rejection based on some selected "maximum" error criteria has more sound statistical basis. For example, one may wish to reject all values outside the range of either 2.5 σ, or 3.0 σ, or 3.29 σ. From Table 4.6 and the accompanying statement concerning "maximum" error, it is seen that this would reject either 1 out of 100, or 3 out of 1,000, or 1 out of 1,000 for those three respective "maximum" errors. For a limited sample, any of these might therefore be used to detect and reject wild values. For most sets it is recommended that rejections not be made using ranges narrower than about 2 σ or 2.5 σ; otherwise truly random and typical observations may be omitted.

As an example to illustrate the application of the above, let us return to Data Set 3 and examine the 23.4 reading. For this set,

$$2.5\,\sigma = \pm\ 3.15'' \quad , \quad 3.0\,\sigma = \pm\ 3.78'' \quad , \quad 3.29\,\sigma = \pm\ 4.15''$$

with a mean of 26.6. The reading of 23.4 would be outside of the desired range if the more conservative 2.5 σ rejection criteria was adopted since it is smaller than 26.6 minus 3.15. If the 3 σ or 3.29 σ criteria was used, the reading would not be rejected since it would fall in these ranges. The high reading of 29.1 would be included in any case. As an exercise, the student may wish to test Data Sets 1 and 2 for possible rejections of wild values.

Once a rejection is made, $\bar{x}$ and σ must be recomputed. The wild value, it must be realized, "contaminated" the original set which will be changed slightly in character after removal of the wild value. Just as removal of a contaminant from the midst of any family should improve it, so should the data set be improved. The improvement, of course, is in the precision index, σ, which will be smaller. Also, the mean will shift away from the value rejected. After removal of the 23.4 reading from Set 3, $\sigma = 1.09''$ and $\bar{x} = 26.74''$ instead of $\sigma = 1.26''$ and $\bar{x} = 26.60''$ as computed before the wild value was rejected.

4-5 THEORY OF RANDOM ERROR PROPAGATION

General Theory

Until now in this chapter, concern has been only with measurements of a *single* quantity. Although several observations were made to comprise a set, the set represented only one quantity, nevertheless. In actual practice, most surveyed quantities are actually derived or computed from more than one single measured quantity. In this sense, they are *indirect measurements*. Even a taped distance is usually not a single observation but the *sum* of several tape lengths and a fractional length. The simple halving of a double angle involves a computation. More complex examples of common indirect measurements are area, plane coordinate, and triangulation computations. All of these employ *direct measurements* of single quantities each of which has its own individual random errors. How the errors accumulate, cancel, decrease, or otherwise behave through the process of computing the final value *indirectly* is termed *propagation of random errors*.

To be able to compute random errors in indirect measurements for all but the simplest situations, the student should have a general knowledge of differential calculus and be able to evaluate partial derivatives. Those without such background will, however, be able to apply the several general formulas to be derived. These emerge as simple algebraic formulas after application of the general formula for propagation of random errors.

General Formula

If

$y = f(x_1, x_2, \ldots, x_n)$ and

$x_1, x_2, \cdots, x_n$ are independently measured, and

$\sigma_{x_1}, \sigma_{x_2}, \cdots \sigma_{x_n}$ are small,

$$\sigma_y = \sqrt{\left(\frac{\partial y}{\partial x_1}\sigma_{x_1}\right)^2 + \left(\frac{\partial y}{\partial x_2}\sigma_{x_2}\right)^2 + \cdots + \left(\frac{\partial y}{\partial x_n}\sigma_{x_n}\right)^2} \tag{4.6}$$

This is the general formula for propagation of random errors. It is used when it is desired to estimate the error in an indirectly measured quantity, y, which is being computed as a function of other measurements, all of which have their individual errors. Errors such as E_{90}, E_{95}, or simply estimates of random errors can be substituted in Equ. 4.6 in place of σ, but the level of certainty of the errors must be consistent throughout the equation. The resulting calculated error for the y value will have the same level of certainty as that of the errors in the x values.

The derivation of Equ. 4.6 is not important for an understanding of its application. Its derivation can be found in some of the references at the end of this chapter.

Applications of the General Formula

Error in a sum. If

$$y = x_1 + x_2 + \cdots + x_n,$$

$$\frac{\partial y}{\partial x_1} = \frac{\partial y}{\partial x_2} = \cdots = \frac{\partial y}{\partial x_n} = 1$$

Substituting these partial derivatives into Equ. 4.6 yields

$$\sigma_y = \sqrt{\sigma_{x_1}^2 + \sigma_{x_2}^2 + \cdots + \sigma_{x_n}^2} \tag{4.7}$$

This is the general formula for *propagation of random errors in a sum*. It could also be shown that the same formula results for *differences*, as well as sums.

Error in a series. If

$$y = x_1 + x_2 + \cdots + x_n \text{ , and}$$
$$\sigma_{x_1} = \sigma_{x_2} = \cdots = \sigma_{x_n} = \sigma_x \text{ (equal precision indexes),}$$

then from Equ. 4.6, $\sigma_y = \sqrt{n\sigma_x^2}$ or

$$\sigma_y = \sigma_x \sqrt{n} \tag{4.8}$$

This is known as the *error in a series* or the *law of compensation*. It is the formula that best illustrates the principle that random errors *tend* to cancel, but never do so completely. In a series, they accumulate in proportion to the square root of the number of opportunities for their occurrence. The student should pause to consider the difference between how random errors accumulate as compared with systematic errors, the latter accumulating in proportion to n rather than $\sqrt{n}$.

Standard error of the mean. The arithmetic mean of a series of repeated measurements of a single quantity is

$$\bar{x} = \frac{x_1 + x_2 + \cdots x_n}{n} = \left(\frac{1}{n}\right)x_1 + \left(\frac{1}{n}\right)x_2 + \cdots + \left(\frac{1}{n}\right) x_n$$

$$\frac{\partial \bar{x}}{\partial x_1} = \frac{\partial \bar{x}}{\partial x_2} = \cdots = \frac{\partial \bar{x}}{\partial x_n} = \frac{\partial \bar{x}}{\partial x} = \frac{1}{n}.$$

If all measurements are made with the same care,

$$\sigma_{x_1} = \sigma_{x_2} = \cdots = \sigma_{x_n} = \sigma_x .$$

Substituting these errors and the partial derivatives into Equ. 4.6 yields

$$\sigma_{\bar{x}} = \sqrt{n\left(\frac{1}{n}\right)^2 \sigma_x^2} \qquad \text{from which}$$

$$\sigma_{\bar{x}} = \frac{\sigma_x}{\sqrt{n}}, \tag{4.5}$$

which is the formula for *standard error of the mean,* introduced in the previous section.

Error in a product. If

$$y = ab, \text{ then } \frac{\partial y}{\partial a} = b \text{ and } \frac{\partial y}{\partial b} = a.$$

Substitutions into Eq. 4.6 yields

$$\sigma_y = \sqrt{a^2 \sigma_b{}^2 + b^2 \sigma_a{}^2} \quad \text{or} \quad \sigma_y = \sqrt{(a\sigma_b)^2 + (b\sigma_a)^2} \tag{4.9}$$

Examples. The next chapters will include detailed sample problems on the applications of the propagation of random error theory. A few simple examples based on the above general formulas will be included here to help explain the general concept of propagation of errors.

Example 4-1. Given: A distance of 600.00 feet was measured in three segments. Different techniques were used for each segment, each of which had been previously tested statistically to determine a value for its standard error. The lengths and their standard errors are as follows:

	Length, ft.	σ_m, ft.
x_1	100.00	± 0.01
x_2	200.00	± 0.05
x_3	300.00	±0.008

What is the standard error in the sum, $x_1 + x_2 + x_3$?

Solution:

$$y = x_1 + x_2 + x_3$$

$$\sigma_y = \sqrt{\sigma_{x_1}{}^2 + \sigma_{x_2}{}^2 + \sigma_{x_3}{}^2} = \sqrt{0.01^2 + 0.05^2 + 0.008^2}$$

$$= \pm\ 0.052 \text{ ft.}$$

The sum and its error can best be expressed as

$$y = 600.0 \pm 0.05 \text{ feet, or possibly } 600.00 \pm 0.05 \text{ feet.}$$

The thousandths place is not significant. It is to be noted that, to the significant figures warranted, the uncertainty in the sum is equal to the uncertainty in the 200 ft. segment. This occurred in this instance because the other two segments were measured with much greater precision. This point serves to indicate that it is often nearly wasted effort to carefully control the errors in one part of an indirect measurement unless consistent control is exercised in other parts. This is an example of the adage "a chain is only as strong as its weakest link."

Example 4-2. Given: A 1,000 meter distance was measured using a tape of 100 meters nominal length. The combined random error from all sources is assumed to be ± 1.0 cm. for each tape length. What is the uncertainty in the total length?

Solution:

$L = n\ell$ where n = number of tape lengths
ℓ = length of tape
L = total length

The uncertainty in the total length is

$$e_L = e_\ell\sqrt{n} = \pm 1.0\sqrt{10} = \pm 3.2 \text{ cm}$$

Therefore, the distance could properly be expressed as 1,000.00 ± 0.03 meters. It is noted that the symbol, e, is used for the uncertainty when it is based on judgment or assumption. Note how the law of compensation works. Even though there was an estimated random error of ±0.01 meters in each of 10 operations, the total error is not expected to be 10 times the separate error. It must be remembered that the actual individual errors could be either plus or minus and thus some cancel each other completely. Furthermore, they will fall in a range between +0.01 and −0.01, so that the accumulated *expected* error must be much less than the conceivable maximum of 0.10 meters. This illustrates the nature of random errors and their tendency to compensate.

Example 4-3. Given: A rectangular tract of land has been measured and its area is to be calculated. The short dimension is 100.00 feet with an estimated uncertainty of ± 0.01 ft. The long dimension is 200.00 feet with an estimated uncertainty of ± 0.03 feet. What is the expected error in the area?

Solution:

$A = \ell w$ where ℓ = length
w = width
A = area

The expected error or uncertainty in the area is

$$e_A = \sqrt{(\ell e_w)^2 + (w e_\ell)^2}$$

$$= \sqrt{(200 \times 0.01)^2 + (100 \times 0.03)^2} = \sqrt{2^2 + 3^2} = \pm 3.6 \text{ sq. ft.}$$

The area would be expressed as

20,000. ± 3.6 sq. ft.

Note that the product has five significant figures according to Rule 2 of Article 4-2. Thus, the uncertainty could be expressed as ± 4 sq. ft. since the units column is the "doubtful" figure. This example is one that can easily be illustrated graphically.

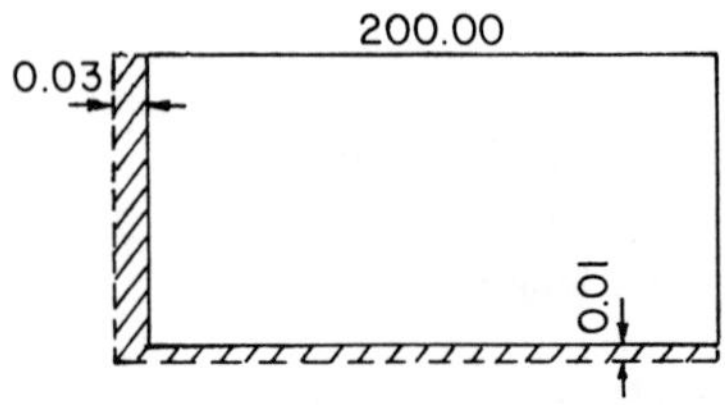

The two narrow cross-hatched portions, representing the area uncertainties due to dimensional uncertainty, have areas of

$$e_w \ell = 0.01 \times 200 = 2 \text{ sq ft. and}$$
$$e_\ell w = 0.03 \times 100 = 3 \text{ sq. ft.}$$

Adding these according to the theory of propagation of random errors in a sum yields

$$e_A = \sqrt{2^2 + 3^2} = \pm 3.6 \text{ sq. ft.}$$

the same answer as when the propagation of random errors in a product was applied.

4-6 PROPAGATION WITHOUT DERIVATIVES

The derivative or differential provides a direct way to derive equations for the change in one variable due to changes in dependent variables. The equations derived for each propagation problem yield an immediate value for the error in the computed quantity caused by estimated errors in all variables used to compute the quantity. Readers without background in calculus may wish to use an alternate method for propagating random errors. Without using calculus, the same answer can be caluculated but the procedure is generally a little longer. The first step in the procedure is to compute the quantity using the "best values" for each variable. Next, each variable is changed by its uncertainty estimate, the quantity is re-computed, and the effects on the quantity are determined. The effects of these changes are computed separately. That is, when one variable is changed by its uncertainty estimate, the others remain unaltered when re-computing the quantity. The separate effects are determined by simply taking the difference each time between the quantity calculated using the best value assumed for each variable and the quantity calculated when the variable is altered by its uncertainty estimate. When altering each variable, it does not matter if it is made larger or smaller by the uncertainty estimate since, if the uncertainties are very small in relation to the variable itself, the absolute value of the effect on the computed quantity will be the same whether the plus or the minus uncertainty is used. The last step is to add the effects (differences) according to propagation in addition theory. The procedure can be illustrated using Example 4-3 as follows:

$$\text{Area} = 100.00 \times 200.00 = 20{,}000 \text{ sq. ft.}$$
$$\text{Area}_{e_1} = 100.01 \times 200.00 = \underline{20{,}002 \text{ sq. ft.}}$$
$$e_1 = 2 \text{ sq. ft.}$$

e_1 = error in area due to error in width

$$\text{Area} = 100.00 \times 200.00 = 20{,}000 \text{ sq. ft.}$$
$$\text{Area}_{e_2} = 100.00 \times 200.03 = \underline{20{,}003 \text{ sq. ft.}}$$
$$e_2 = 3 \text{ sq. ft.}$$

e_2 = error in area due to error in length

$$e_A = \sqrt{2^2 + 3^2} = \pm\ 3.6 \text{ sq. ft.}$$

For most problems, calculus gives a more immediate solution once the error equation has been derived, but the above "rate of change" method will yield the same results and is suitable when partial derivatives become cumbersome or when the analyst lacks the math background to perform the calculus operations. Its disadvantage is the need for more computational steps beyond that needed to solve a single error equation. This disadvantage is not particularly great for problems with only a few variables when an electronic calculator is used.

For the sake of encouraging use of the more sophisticated and efficient method, the approach employing calculus will be emphasized throughout the book, but solutions will occasionally be given for examples using both approaches. Readers without calculus background may wish to try solving the text examples by the method introduced in this section.

4-7 FROM THEORY TO PRACTICAL ANALYSIS

The qualified surveying technician is expert in *execution* of survey measurements. The professional surveyor is expert in *analyzing* survey measurements, as well as in techniques of executing measurements. The technician must understand much of the analysis portion in order to make accurate and precise measurements.

Chapters 5 through 8 are devoted to analyzing measurements, both in terms of systematic and random errors. The focus is on the latter. The organization of the chapters is based on typical surveying measurement situations, using equipment described in Chapter 3. The following chapters explain, through practical examples, the instruments and their limitations as well as the theories developed in this chapter. After study of Chapters 4 through 8, along with review and memorization of basic definitions explained earlier in the book, the reader should be able to analyze most common surveying measurement problems and hopefully also extend this knowledge to the solution of many other measurement situations not covered by examples. The student should not attempt to memorize specific equations and approaches, but instead, try to see how the theories and general formulas are being applied. Actual situations encountered in practice will undoubtedly vary from the "typical" examples given.

Problems - Chapter 4

4.1 It is common knowledge that the sun is 93,000,000 miles from the earth. Can the uncertainty of this distance be estimated from the number as expressed? Why or why not?

4.2 How many significant figures does the measurement 40.6020×10^{-3} have?

4.3 Express the following answers to the correct number of significant figures:

a) (36.4 + 13.26 − 51.480)
b) (363,481.8 ÷ 46.21)
c) Conversion of 1392.86 feet to meters
d) 686.892 sin 30°
e) Area of ⅓ of a circle with radius 100.00 meters

4.4 Express the following answers to the correct number of significant figures:

a) 0.0254 × 368.60
b) (468.859 sin 32° 12′) ÷ sin 64° 18′
c) $32\pi + 32.21$
d) Conversion of 871,212 sq. ft. to acres
e) Conversion of 36.867 degrees to degrees, minutes, seconds.

4.5 What is the relative error implied by a distance measurement stated as 1691.833 meters?

4.6 Compute the standard deviation of a single value using only the first 15 readings of Data Set 1 of the text examples.

4.7 Using the results of Data Set 2 for σ, what would be the expected standard error of the mean of 6 repetitions using this method in the future?

4.8 In a data set of repeated measurements yielding $\sigma = \pm 3.8$ mm, compute the following:

a) E_{95} b) E_{99} c) E_{50}

4.9 If the data set of Problem 4.8 contained 200 repetitions, how many readings would be expected to lie in the range ± 6.25 mm. from the mean?

4.10 The following readings were recorded for a theodolite error test.

11.1	14.2	18.1	9.8	10.5
13.4	14.5	15.5	9.0	17.8
14.2	12.2	15.8	13.4	13.2
14.6	14.2	9.5	12.6	15.4
13.4	12.0	10.2	9.8	12.5

For the above, compute the standard deviation, σ, and plot a histogram.

4.11 Test the data in Problem 4.10 for wild values using 2.5 σ as rejection criteria.

4.12 Test the data in Problem 4.10 for randomness and symmetry by comparing the mean, median, and mode.

4.13 Compute E_{90} and E_{95} for the data in Problem 4.10, count the number of values within these ranges, as well as the σ range and compare this count with the number expected from theory.

4.14 Using the usual convention for significant figures as a precision index, compute the uncertainty in the sum 36.4 + 13.26 − 51.480.

4.15 Three distance segments, each having uncertainties of ±1.0 cm. were added to four others, each having uncertainties of ±0.5 inches. What is the uncertainty in the sum of these seven segments, in feet?

4.16 What is the uncertainty in the radius of a circle, given the circumference as 200.00 feet ± 0.05 feet?

Additional Reading References – Chapter 4

1. Barry, Austin B. *Errors in Practical Measurement in Science, Engineering, and Technology.* New York: John Wiley & Sons, Inc., 1978, Chapter 3-7, Appendix A and B.

2. Brinker, Russell C. and Wolf, Paul R. *Elementary Surveying.* Sixth Ed. New York: Harper and Row, 1977, Chapter 2.

3. Mikhail, Edward M. and Gracie, Gordon. *Analysis and Adjustment of Survey Measurements.* New York: Van Nostrand Reinhold Company, 1980, Chapters 1-2.

4. Moffit, Francis H. and Bouchard, Harry. *Surveying.* Seventh Ed. New York: Harper and Row, 1981, Chapter 4.

5. Mueller, Ivan I. and Ramsayer, Karl H. *Introduction to Surveying.* New York: Frederick Ungar Publishing Company, 1979, Chapter 2.

6. Schmidt, Milton O. and Rayner, William H. *Fundamentals of Surveying.* Second Ed. New York: D. Van Nostrand Co., 1978, Art. 7-5, 7-6, Chapter 8.

Section B
ANALYZING MEASUREMENTS

CHAPTER 5

ANALYSIS OF DISTANCE MEASUREMENTS

5-1 ANALYSIS OF TAPED MEASUREMENTS

Importance of Understanding Taping

Tapes and taping have been introduced in Chapter 3. Details of how to unreel and make readings with a steel tape are not subject matter for this book. Training in these skills is best achieved through field exercises or on-the-job training and can be developed easily by any reasonably adept individual. Although proper manipulation of the tape itself is important, more important is an understanding of the several sources of errors in taping. Many technicians, highly skilled in physically manipulating the tape so as to achieve good precision, fail in achieving accuracy because they misunderstand the error sources and how to deal with them. Both from the standpoint of physical use of the tape, and full appreciation of error sources, taping probably requires more skill than most other surveying measurements because the quality of results are so much at the mercy of the observer. Other instruments have built-in automation to some degree such that even untrained individuals could probably achieve accurate results a high percentage of the time. This is not so with the tape where measurements are highly susceptible to several personal and natural errors and mistakes. It seems inappropriate that beginning field technicians are often given the job of rear tapeman. It might be more appropriate that they learn the use of automated equipment first, then graduate to tapeman after a better appreciation of measurement has been achieved.

Systematic Errors

The length of the tape is affected by (1) manufacture, (2) temperature, and (3) tension. A horizontal distance is also affected by any differences in elevation along the tape and the amount of support the tape has between its ends. The first of these is an instrumental error; the others are natural errors. All of these error sources are systematic and thus, corrections can be made for their effect. The following discussion will be in the context of corrections, keeping in mind that each is a correction for a systematic error. The various corrections are as follows.

Length correction. The 100-foot surveyor's tape is generally manufactured to its nominal length at either 68°F or 72°F and at tensions ranging from 10 to 30 pounds. The calibration information may or may not be furnished with the tape. A tape can be calibrated by comparing its length with standards established for that purpose. The tension, temperature, and support condition (usually either fully supported or end supported) must be read, along with the length. The amount that the tape varies from nominal is called the calibration or *length correction,* C_ℓ. For example, a tape might be found to be 100.008 feet long at 70°F, 15 lbs pull, fully supported. $C_{\ell_1} = +0.008$ under these conditions.

Temperature correction. The length of a steel tape changes with temperature. It becomes longer as it gets warmer according to the formula

$$C_{t_1} = K_t\ (t_f - t_s)\ell\ , \tag{5.1}$$

where

C_{t_1} = temperature correction, one tape length

K_t = coefficient of thermal expansion

t_f = field temperature

t_s = temperature at which the tape was standardized

ℓ = the nominal length of the tape.

Values for the coefficient of thermal expansion of steel were given in Section 3-4. In the formula as stated, t_f and t_s must have units consistent with that of K_t, be it Fahrenheit or Celcius.

Tension correction. Steel is an elastic material within the range of tension possible by human power alone. Its stretching characteristics are therefore essentially linear. (See Figure 2.1) The elasticity of any material is defined by its modulus of elasticity

$$E_m = \frac{s}{\epsilon}$$

where

s = stress in lbs/in², (or Kg/cm², etc.),

ϵ = unit strain.

For a steel tape,

$$s = \frac{P}{A}\ ,$$

$$\epsilon = \frac{e}{\ell}$$

where

P = pull (tension)

A = cross sectional area of the tape

e = tape's elongation under pull, P

ℓ = nominal tape length.

We are concerned with the difference in the length of a tape correction if a tension other than that used for standardizing is applied. From the above it can be shown that

$$C_{p_1} = \frac{(P_f - P_s)\ell}{AE_m} \tag{5.2}$$

where

P_f = field tension

P_s = tension applied for standardization

C_{p_1} = correction for tension (elongation, e, for $P_f - P_s$)

E_m = modulus of elasticity for steel (see Section 3-4).

Using average values for thickness and width of a tape as stated in Section 3-4, it could be shown that tapes are about 0.005 in^2 to 0.006 in^2 in cross-sectional area.

Slope correction. If the tape is not horizontal a correction must be applied for slope. This is required except when taping over very flat and uniform terrain or when the tape ends can be held at the same elevation and the tape suspended between. An observation of the slope must be made in the form of either a vertical angle or elevation difference. Referring to Fig. 5.1,

$$h = s + C_g \tag{5.3}$$

and

$$C_{g_1} = -s(1-\cos\gamma).$$

where

h = horizontal length

s = length of uniform slope, one tape length

C_{g_1} = slope correction, one tape length

γ = vertical angle

It will be noted that the slope correction is always negative.

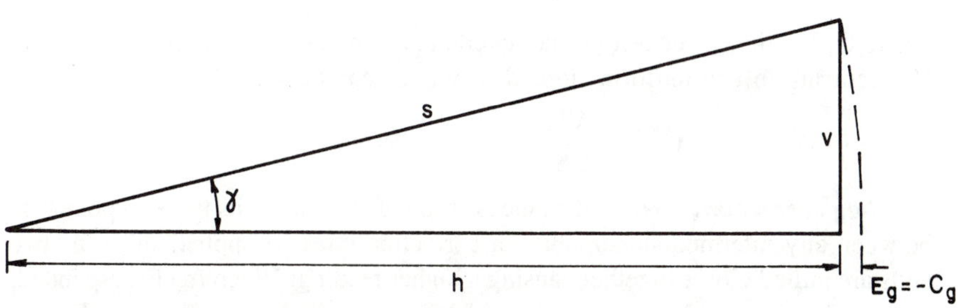

Figure 5.1 Slope Correction

If the elevation difference, v, is measured instead of the vertical angle, an expression of the correction can be derived as follows:

$$E_g = s-h = s-(s^2-v^2)^{½} = s-s(1-\frac{v^2}{s^2})^{½}$$

where

$$v = \text{difference in elevation}$$
$$E_g = \text{error in reading due to slope.}$$

Expanding the term $(1-\frac{v^2}{s^2})^{½}$ yields

$$1 - \frac{v^2}{2s} - \frac{v^4}{8s^3} - \frac{v^6}{16s^5} - \cdots\cdot$$

Substituting this series expansion into the above expression yields

$$E_g = s-s\,(1 - \frac{v^2}{2s} - \frac{v^4}{8s^3} - \frac{v^6}{16s^5} - \cdots)$$

$$E_g = + \frac{v^2}{2s} + \frac{v^4}{8s^3} \cdots$$

For ordinary taping, up to slopes of about 10%, only the first term of this equation is necessary, in which case,

$$C_{g_1} = - \frac{v^2}{2s}\,. \tag{5.4a}$$

Occasionally, it may be necessary to apply the second term for very steep slopes or for very precise taping.

If a slope is uniform over several tape lengths, n, this equation can be written

$$C_g = - n\frac{v^2}{2s} \tag{5.4b}$$

where C_g = the correction for the several tape lengths. Or, the vertical distance, V, occurring in any uniform slope distance, S, can be solved by

$$C_g = - \frac{V^2}{2S} \tag{5.4c}$$

Sag correction. When the tape is suspended between its two end points, or between any intermediate supports, a correction must be applied since the two ends are pulled closer together causing a higher reading. When freely suspended, the tape forms an arc approximating either a catenary curve or parabola. For the purpose of deriving a correction formula, either can be assumed. Let us assume that the tape hangs in a parabola ACB as shown in Fig. 5.2. The equation of this arc is

$$y = Ax^2 \quad \text{and its slope} \frac{dy}{dx} = 2Ax.$$

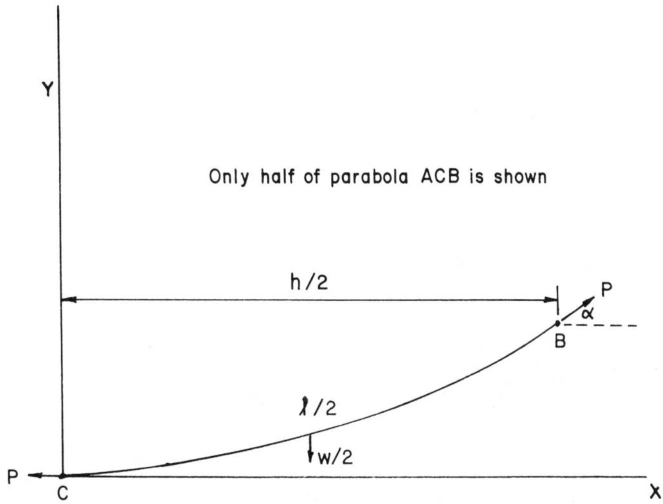

Figure 5.2 Sag Correction

In this figure

w = weight of tape between supports A and B
h = horizontal length between supports A and B
P = tension on the tape
ℓ = length of tape between supports A and B.

From statics, the two component horizontal and vertical forces are

$$H = P \cos \alpha$$

and

$$V = P \sin \alpha = \frac{w}{2}$$

from which

$$\tan \alpha = \frac{\sin \alpha}{\cos \alpha} = \frac{w}{2H}$$

At B ,

$$\tan \alpha = \frac{dy}{dx} = 2A\frac{h}{2} = Ah$$

from which

$$A = \frac{\tan \alpha}{h} = \frac{w}{2hH}$$

and

$$\frac{dy}{dx} = 2\,\frac{wx}{2hH} = \frac{wx}{hH}$$

From the calculus, $s = \int \sqrt{1 + \left(\frac{dy}{dx}\right)^2}\, dx$,

from which $\frac{\ell}{2} = \int_o^{h/2} \sqrt{1 + \left(\frac{wx}{hH}\right)^2}\, dx$

Expanding $\sqrt{1 + \left(\frac{wx}{hH}\right)^2}$ by the binomial series yields

$$\ell = 2 \int_o^{h/2} [1 + \frac{w^2\,x^2}{2h^2 2H^2} - \frac{w^4\,x^4}{8h^4 8H^4} + \cdots]dx$$

Integration yields

$$\ell = 2[\frac{h}{2} + \frac{w^2 h}{48H^2}] \text{ for the first two terms.}$$

Therefore, $\ell = h + \frac{w^2 h}{24H^2}$.

And , as H approaches P and h approaches ℓ , the error due to sag is

$$\ell - h = \frac{w^2 \ell}{24P^2} = E_s$$

Therefore, the correction due to sag, for a tape length supported at two end points is

$$C_{s_1} = -\frac{w^2 \ell}{24P^2} \qquad (5.5)$$

This formula gives the correction for any portion of the tape which is suspended. It is emphasized that the *weight, w , and length,* ℓ *, is that between the portion suspended.* Note that the sag correction is always negative.

Alignment. The shortest distance between two points is a straight line. If the tape is laid along any other line, readings will be too high. The alignment error is one which is random in nature but systematic in effect. That is, the tapemen try to align the ends along a straight line, but have a random error to either side of the line. But, whether to the right or the left of the line, the effect is always the same in sign. A lateral displacement from the true line would cause a systematic error

$$E_{a_1} = \frac{d_a}{2\ell}, \tag{5.6}$$

where

E_{a_1} = alignment error, one tape length
d_a = lateral displacement
ℓ = length of tape

The derivation of this equation is the same as that for the slope correction, Equation 5.4.

If alignment is accomplished within 0.5 feet, which is possible by eye, the error would be about 0.001 ft. which is nearly negligible, but nevertheless cumulative over long distances. If alignment is done with a theodolite, the error is quite negligible. Some surveyors make a correction based on an estimate of d_a; hence,

$$C_a = -n\,E_{a_1}$$

where n = number of tape lengths,
C_a = alignment correction.

Counteracting Systematic Errors

Some corrections can be avoided by observing in such a way that the correction is zero. The tension correction and sometimes the slope and sag corrections can be handled in this manner. If standard tension is used, there is no correction for tension. If the slope is measured and the error calculated for a tape length, this amount can be added physically in the field using an ''add'' tape. Sometimes, a calculation can be made to determine the tension required to cause the tape to be its nominal length when suspended. If this tension is not so high as to be impractical to apply, it can be used and the sag correction is thus counteracted. Usually, however, corrections must be applied to data as read under uncontrollable, but observable conditions.

Calculating Corrections for Systematic Errors

Taping corrections are made using the familiar formula

$$T = R + C_T \quad \text{or} \quad R = T - C_T \tag{2.4}$$

as discussed in Section 2-6. The first equation is used to solve for the true distance between two established points, the reading between which has been observed. The second equation is used to solve for the reading necessary to lay out or establish a desired (true) distance from one established point.

It is usually most efficient and least susceptible to making mistakes if each error source is individually considered and a total correction then calculated by algebraic addition of the corrections from each source. In some instances, the total effect for the full distance is the effect in one tape length multiplied by the number of tape lengths. But this is not always the case and each situation must be analyzed individually. In any case the basic formulas above should be strictly employed since doing so allows the algebraic signs to work themselves out without necessity for the useless mental exercise of always having to stop and consider whether readings are "too long", or "too short". The latter approach, rather than the recommended one, is a mark of a surveyor who has not learned basic algebra. It is a bad habit which wastes time and mental energy and invites mistakes.

Several examples will illustrate the application of taping corrections.

Example 5-1 Given: A 50 meter steel tape has been used to measure a distance. The observed reading is 261.450 m. The tape is known to be 50.001 meters long at 5 Kg. tension and 20°C, fully supported throughout its length. The field temperature is 20°C. Compute the true distance.

Solution: Note that the distance was observed by laying out 5 full tape lengths and then reading 11.450 m from the last pin set to the terminal point. Since there was nothing mentioned in the problem statement about field tension, it must be assumed that the standardization tension was used, thus avoiding a tension correction. Since neither slope or sag is mentioned, it must be assumed that the tape was used fully supported on flat terrain and so there is no sag or slope correction. Obviously, the only correction necessary is for calibrated length. The calibrated length is the *true* length of the tape and 50.000 is what is *read* from the instrument between its end marks.

$$\ell_T = \ell_R + C_{\ell_1}$$

where

$$\ell_T = 50.001 \text{ m}$$
$$\ell_R = 50.000 \text{ m}$$

from which $C_{\ell_1} = +0.001$ m. This correction is the same for every full tape length and is assumed to be distributed uniformly throughout the tape. The number of tape lengths is

$$n = \frac{261.45}{50} = 5.2290 .$$

The correction for the full distance is

$$C_\ell = nC_{\ell_1} = 5.2\,(+0.001) = +0.005 \text{ m.}$$

and

$$L_T = L_R + C_\ell = 261.450 + 0.005 = 261.455 \text{ m.}$$

where

$$L_T = \text{true length of the line}$$
$$L_R = \text{observed reading.}$$

Note that the significant figures in C_ℓ are ultimately controlled by the rule of addition applied to the final computation in solving for L_T and thus n need only have two significant figures, the extra one being merely to avoid round-off error in solving for C_ℓ. The student should carefully consider this concept in the remaining examples in order to learn to avoid carrying unnecessary digits in computations.

Example 5-2. Given: The same tape and observed length of Example 5-1 but with a field temperature of 28°C. Compute the true distance.

Solution: This change in conditions adds only the temperature correction. It must be assumed that the temperature remained the same for all tape lengths. Therefore, the correction per unit of length is constant, and

$$C_t = nK_t\ (t_f - t_s)\ \ell$$

where

$$n = \text{number of tape lengths} = 5.2$$
$$K_t = 0.0000116 \text{ m/m/°C (from Section 3–4)}$$
$$t_f = 28°C$$
$$t_s = 20°C$$
$$\ell = 50 \text{ m.}$$

from which

$$C_t = +0.024 \text{ m.}$$

Note again, from the standpoint of significant figures, it does not matter that 50 was used for ℓ instead of the actual value of 50.001 since there are only two significant figures in the final answer.

It is at this point that many surveyors forget that there is still a calibration length correction. It has not disappeared merely because the weather changed. The total correction is, therefore,

$$C_T = C_\ell + C_t = +0.005 + 0.024 = 0.029 \text{ m.}$$

and

$$L_T = 261.450 + 0.029 = 261.479 \text{ m.}$$

It should be noted that, in this series of examples, the actual position of the two ground points is not the same since the true distance is changing each time the sample is altered. This might be confusing unless it is realized that the same observed reading is used in all the examples merely for convenience in illustrating the calculations. The position of the ground points should be considered unique in each example.

Example 5-3. Given: The same tape, observed reading and field conditions in Example 5-2 except that the tape was used with the ends held at the same elevation, in the fully suspended condition under 10 Kg. tension. The tape weighs 1.0 Kg. Its cross sectional area is 0.035 sq.cm. and modulus of elasticity is 2,050,000 Kg/cm². Find the true distance.

Solution: Analysis of this situation indicates that there will be a sag correction and a tension correction. It must be assumed that the tape was suspended 5 times between its 0 and 50 m. marks and one time through a length of 11.450 m. The sag correction for each of the 5 full lengths will be the same. The correction for the fractional length must be a separate computation since the corrections are not only directly proportional to length but also to the square of the weight between supports. For the 5 full lengths,

$$C_{s_5} = -n \frac{w_1^2 \ell}{24\ P^2}$$

where

$$n = 5$$

$$w_1 = 1.0 \text{ Kg}$$

$$\ell = 50 \text{ m}$$

$$P = 10 \text{ Kg}$$

from which

$$C_{s_5} = -0.1042 \text{ m}$$

For the fractional tape length,

$$C_{s_f} = -\frac{w_f^2 \ell_f}{24\ P^2}$$

where

$$w_f = \text{weight of 11.450 meters of tape} = 0.229 \text{ Kg}$$

$$\ell_f = 11.45 \text{ m}$$

$$P = 10 \text{ Kg.}$$

from which

$$C_{s_f} = -0.0003 \text{ m}$$

$$C_s = C_{s_5} + C_{s_f} = -0.1045 \text{ or } -0.105 \text{ m.}$$

Next, we must account for the extra 5 Kg. of tension beyond the standardization tension. The correction for this remains the same per unit length and thus the correction for one tape length can be multiplied by the number of tape lengths.

$$C_p = n \frac{(P_f - P_s)\ell}{AE_m}$$

where

$$n = 5.22$$

$$P_f = 10 \text{ Kg.}$$

$$P_s = 5 \text{ Kg.}$$

$$\ell = 50$$

$$A = 0.035 \text{ sq. cm.}$$

$$E_m = 2{,}050{,}000 \text{ Kg/cm}^2$$

from which

$$C_p = +0.0182 \text{ m.}$$

It should be noted that the tape could have been considered as one continuous length of $\ell = 261$ meters instead of 5.22 lengths of 50 meters and the same answer would have resulted.

$$C_T = C_\ell + C_t + C_s + C_p$$

$$= +0.0050 + 0.0242 - 0.1045 + 0.0182 \text{ m.}$$

$$= -0.057 \text{ m.}$$

and

$$L_T = 261.450 - 0.057 = 261.393 \text{ m.}$$

Although the final answer for C_T has only three decimal places, four were used in the intermediate computations to avoid round-off errors.

Example 5-4. Given: The same tape, observed reading and field conditions of Example 5-2 except that the tape was fully supported on an average slope of 2.5%. Find the true distance.

Solution: The results of Example 5-3 can, of course, be ignored. The only corrections to be considered are calibration length, temperature, and slope. Since there is a uniform slope throughout the length, the correction for one tape length can be multiplied times the number of tape lengths.

$$C_g = -n \frac{v^2}{2s}$$

where

$$n = 5.22$$

$$v = \text{elevation difference for 1 tape length} = 1.25 \text{ m}$$

$$s = \text{one tape length} = 50 \text{ m.}$$

from which

$$C_g = -0.0816 \text{ m.}$$

$$C_T = +0.0050 + 0.0242 - 0.0816 \text{ m} = -0.052\text{m.}$$

and

$$L_T = 261.450 - 0.052 \text{ m} = 261.398 \text{ m.}$$

Example 5-5. Given: A distance was taped along the edge of a highway. The tape of Example 5-1 was used. The temperature was 0°C. Since the grade was very uniform and the line of sight was clear, the zenith angle was measured with a theodolite. This angle was 90° 46′ 20″. The reading with the tape fully supported was 1,265.89 m. Find the true distance.

Solution:

$$n = 1{,}265.89/50 = 25.31$$

$$C_\ell = nC_{\ell_1} = 25.31\,(+0.001) = +0.0253 \text{ m.}$$

$$C_t = nK_t(t_f - t_s)\ell = 25.31\ (0.0000116)(-20)(50)$$

$$= -0.2936 \text{ m}$$

$$C_g = -S(1 - \cos\gamma)$$

where

$$S = \text{slope distance} = 1265.89$$

$$\gamma = \text{vertical angle} = 90° - (90°46'20'') = -0°46'20''$$

$$C_g = -\ 1265.89\ (0.00009082) = -0.1150 \text{ m.}$$

$$C_T = C_\ell + C_t + C_g$$

$$= +0.0253 - 0.2936 - 0.1150 \text{ m}$$

$$= -0.3883 \text{ m}$$

and

$$L_T = 1{,}265.89 - 0.38 = 1{,}265.51 \text{ m}$$

It should be noted that the true horizontal distance, L_T, could have been calculated from the corrected slope distance, L_s, as follows:

$$L_s = 1{,}265.89 + 0.025 - 0.294 = 1265.621 \text{ m}$$

and

$$L_T = 1265.621 \cos(-0°46'20'') = 1265.51 \text{ m}$$

Example 5-6. Given: A calibrated steel tape, which is 99.983 feet long between the 0 and 100-ft. marks at 68°F, 15 lbs. tension, fully supported. It is desired to establish a distance of one mile for a bicycle wheel calibration course to be used for TAC/RRCA certification of road races. The average field temperature is 45°F. What should be read from the tape to establish the distance?

Solution: First, it must be recognized that this is a problem in solving for a reading, given the true distance. But, the corrections and their signs remain unchanged. The principal difference in this and previous examples appears only in the last step.

$$n = 5280/100 = 52.8$$

$$C_{\ell_1} = \ell_T - \ell_R = 99.983 - 100.000 = -0.017 \text{ ft.}$$

$$C_\ell = nC_\ell = 52.8\ (-0.017) = -0.898 \text{ ft.}$$

$$\begin{aligned} C_t &= nK_t\ (t_f - t_s)\ell \\ &= 52.8(0.00000645 \text{ ft/ft/°F})\ (45° - 68°\text{F})\ (100) \\ &= -0.783 \text{ ft.} \end{aligned}$$

$$C_T = C_\ell + C_t = -0.898 - 0.783 = -1.681 \text{ ft.}$$

$$L_R = L_T - C_T = 5280.00 - (-1.681) = 5281.68 \text{ ft.}$$

Note the strict attention to algebraic signs and the simplicity of the solution when the formula relating the true value to the reading and total correction is used in its proper form. The mental exercise of deciding whether to add or subtract the 1.68 ft. is avoided with this approach. As a mental check, however, it can be reasoned that some extra amount must be added to the nominal reading since the tape was short when calibrated and then shortened even more when it got colder. A reading of precisely 5280.00 would have been 1.68 ft. short of the desired amount. In the field, this distance of one mile would be established by laying off 52 tape lengths, placing pins at the 100-ft. mark for each tape length, then making the final mark at a reading of 81.68 ft. beyond the last pin set.

Random Errors

There are random errors in the determination of each of the systematic corrections illustrated above. For example, the temperature is a variable and has an uncertainty. The effect of the uncertainties in each of these variables on the length can be estimated by applying the general formula for propagation of random errors to each respective correction formula. The uncertainties in the variables are usually mere estimates based on judgment and a general knowledge of the reading precision and other characteristics of the various devices or methods used to measure the variables. Whether such uncertainties are mere estimates or based on statistical analyses, if they are reasonable approximations, the process of propagating them through the measurement is a reliable means of estimating the *expected error* in the distance. Although there are probably additional small random errors caused by undetected personal, natural, and instrumental sources, the primary random errors in taping are as follows:

Reading and marking. This error is one of the principal personal errors in taping. It is comprised of random errors in: (1) setting pins at the zero mark, (2) aligning the rear mark of the tape at the previously set pins and the beginning point, (3) aligning an even unit mark at the last pin set for measurement of the last fractional distance, (4) estimating the reading at the forward end of the tape at the terminal point. This error is mostly one of setting and aligning pins on tape marks. It is a function of the ability of the observers to estimate the center of a pin, among other things. Only one reading is actually made–at the end of the line. The error is influenced by many factors–roughness of terrain, extent of plumb bob use, straightness of pins (even taping pins are instruments), ground hardness, etc. The simplest way to analyze this error is to estimate a reasonable value for the average uncertainty of one occurrence of aligning, setting, and reading, and to assume this error is the same for all occurrences. This error occurs *two* times for each full *or* fractional tape length. It follows the law of compensation. Therefore,

$$e_{r_1} = \pm e_r\sqrt{2} \tag{5.7}$$

where

e_r = error in aligning and setting pins, marking, or reading one tape end

e_{r_1} = reading and marking error for one full or fractional tape length.

For n tape lengths,

$$e_{L_r} = \pm e_{r_1}\sqrt{n} = \pm e_r\sqrt{2n} \tag{5.8}$$

where

e_{L_r} = reading and marking error for a distance comprised of n tape lengths.

Temperature error. In the temperature correction formula, let us assume K_t, ℓ, and t_s are constant or without error. The variable, therefore, is t_f or Δt. Applying the general formula for error propagation to C_t,

$$\frac{\partial C_t}{\partial \Delta t} = \pm K_t \ell$$

and

$$e_{t_l} = \sqrt{(K_t \ell)^2 e_t^2}$$

from which

$$e_{t_l} = \pm K_t \ell e_t \tag{5.9}$$

where

e_{t_l} = error in one tape length, ℓ, due to e_t
e_t = error in determining temperature for that tape length.

If the temperature is observed each tape length, and the error in estimating temperature is the same,

$$e_{L_t} = \pm K_t \ell e_t \sqrt{n} \tag{5.10}$$

where

e_{L_t} = error in total length comprised of n tape lengths of length ℓ.

But, we must be very careful in this analysis. The temperature is usually *not* read each time the tape is advanced, but is instead read once or a suitable average used for the entire line. Thus, the error does not tend to compensate according to the law of compensation but accumulates directly with n. Or,

$$e_{L_t} = \pm K_t \ell e_t n$$

from which

$$e_{L_t} = \pm K_t L e_t \tag{5.11}$$

where

L = the total line length.

This is the formula which will be applied in most cases. Note that it could have been derived directly by using L in the formula for C_t. It should be noted that the derivation could have been simplified merely by taking the derivative of C_t with respect to Δt and then calling the differentials the errors, instead of applying the longer general formula for error propagation. That is

$$\frac{dC_t}{d\Delta_t} = \pm K_t \ell$$

$$dCt = \pm K_t \ \ell \ d\Delta T$$

Substituting e_{t_1} and e_t for dC_t and $d\Delta T$ yields Equ. 5-9.

Tension error. The explanations will be abbreviated somewhat for the remainder of the derivations, it being assumed that the student is beginning to grasp the application of error propagation theory and the presence of random errors in all quantities read. In the tension correction formula, the field tension P_f will be the only variable considered, although the error in the standarization tension, P_s, and in the cross-sectional area, A, should not be overlooked. Applying propagation theory,

$$e_{p_1} = \pm \left(\frac{\ell}{AE_m}\right) e_p \tag{5.12}$$

where

e_{p_1} = error in one tape length due to e_p

e_p = error in determining tension for one tape length.

Unlike temperature, the tension must be read or estimated (an estimate is actually a reading) for each tape length; therefore, the errors propagate according to the law of compensation, and

$$e_{L_p} = \pm \left(\frac{\ell}{AE_m}\right) e_p \sqrt{n} \tag{5.13}$$

where

e_{L_p} = error in total length

n = number of tape lengths (actually, number of times tension is applied).

This equation assumes that the error, e_p , is the same for each tape length. All of the above assumes that the tension measuring device, mechanical or human, was calibrated with no tendency to read high or low.

Slope error. If a vertical angle is measured for a tape length, ℓ, the error in the respective distance, from application of the propagation of error formula to Equ. 5.3 and 5.4 is

$$e_{g_1} = \pm \ \ell \sin \gamma \ e_\gamma \tag{5.14}$$

where

$$\ell = \text{one tape length (slope distance)}$$

$$e_\gamma = \text{error in vertical angle, } \gamma.$$

or, if the vertical distance between the two tape ends is measured,

$$e_{g_1} = \pm \frac{v}{\ell} e_v \tag{5.15}$$

where

$$v = \text{difference in elevation between tape ends}$$
$$e_v = \text{error in elevation difference, v.}$$

If the angle or slope is measured over the full length, L, (only occurs if slope is uniform over length, L),

$$e_{L_g} = \pm L \sin \gamma\, e_\gamma \tag{5.16}$$

or

$$e_{L_g} = \pm \frac{V}{L} e_v \tag{5.17}$$

where

$$e_{L_g} = \text{error in total length due to uncertainty in vertical angle or slope.}$$

$$V = \text{difference in elevation between the two ends of the line.}$$

The usual case in taping is to measure a percent slope for each tape length, since terrain changes constantly. Since a reading occurs each time, and since the *slope* itself (v/ℓ) changes, the errors propagate according to the rule for addition of random errors. Thus

$$e_{L_g} = \pm \sqrt{(g_1 e_{v_1})^2 + (g_2 e_{v_2})^2 + \cdots + (g_n e_{v_n})^2} \tag{5.18}$$

where

$$g_1, g_2, \cdots g_n = \text{the slope, } v/\ell \text{ for the particular length, } \ell_1, \ell_2, \text{ etc.}$$

Note the variation in error in length due to the slope itself. Given a constant error in estimating v, the error in length increases directly with slope.

Sag error. Sag varies with tension on the tape. Everything else might be considered constant in the sag formula, although the weight uncertainty could be considered, especially if a precise scale was not available. Applying the general propagation formula to C_s, taking the derivative with respect to tension, the result is

$$e_{s_1} = \pm \frac{w^2 \ell}{12P^3} e_p \tag{5.19}$$

where

e_{s_1} = error in one tape length, ℓ,
e_p = error in tension for one tape length.

If the same tension and the same care in applying tension is used for each tape length,

$$e_{L_s} = \pm \frac{w^2 \ell}{12P^3} e_p \sqrt{n} \tag{5.20}$$

where

e_{L_s} = the error due to sag for the total length
n = number of full tape lengths.

A complete analysis would necessitate adding an error for the last fractional tape length, substituting appropriate values for w and ℓ into Equ. 5.19, then adding this result to the result from Equ. 5.20 using random errors in addition procedures. This small amount is probably neglibible and can ordinarily be neglected.

It is to be noted that errors due to sag can be decreased considerably simply by applying greater tension, since the error decreases as the third power of the tension. This is one reason why it is recommended to apply more tension when suspending the tape than when using it fully supported. Up to a point where limitations due to fatigue or muscular strength start to increase e_p, as much tension as possible should be applied when the tape is suspended. To keep e_p minimized, a tension handle should be used. As will be seen in examples to follow, this error must be carefully controlled. Otherwise, it can become the largest contributor to uncertainties in taped distances.

Alignment error. The alignment error is peculiar in that it is random in nature but systematic in effect. That is, even though setting a taping pin on line is random with respect to the line, the reading will always be too high whether the pin is right or left of the line. Therefore it would be improper to treat this error in the same family of random errors explained above. The best thing a surveyor can do with alignment errors is apply a small negative correction, as explained on page 127, if deemed necessary, then consider the random error in this estimate as negligible. This will be the case if ordinary alignment procedures are employed.

Calibration error. It would be wrong to assume that the tape's calibrated length is an absolute value. There is some error in the process of calibration and even in the calibration line itself. It is expected that calibration errors are small, but they should be considered, nevertheless. All that can be done practically is to estimate an uncertainty in the calibrated length as determined. If it is estimated to be e_ℓ, for example, the total uncertainty in a line's length from this source would be

$$e_{L_\ell} = \pm\, e_\ell n \tag{5.21}$$

where

$$n = \text{number of tape lengths.}$$

It is noted that this error would not tend to compensate and therefore must be multiplied by n rather than $\sqrt{n}$. Whatever its value, it is the same with each tape length.

Propagation of Random Errors in Taping

The process of estimating uncertainty in a taped distance is not difficult when the errors are calculated separately from each error source as explained above. Once the effect on the line's length from each source is determined, the total error is simply the square root of the sum of the squares of the separate errors (law of addition of random errors). The calculations can become a little cluttered if lines are long and many changes in conditions (temperature, slope, support condition, etc.) occur, necessitating making separate calculations for various portions of the line. A few examples will illustrate the application of the formulas.

Example 5-7. Given: A party chief estimates that both he and the rear tape operator can read and mark each tape end to ± 0.01 ft. using a 100-ft. steel tape. What is the uncertainty in a 1000-ft. distance laid out with this tape due to this error source?

Solution:

$$e_{L_r} = \pm\, e_r\sqrt{2n} \tag{5.8}$$

where

$$e_r = \pm\, 0.01$$
$$n = \text{number of tape lengths} = 10$$

from which

$$e_{L_r} = \pm\, 0.045 \text{ ft.}$$

Example 5-8. Given: The same tape, total distance, and error of Example 5-7 and, in addition, an error estimate of ± 5°F in the average temperature for the total distance and an error estimate of ± 2 lb. in reading tension from a tension handle. What is the total expected uncertainty in the line from the error sources mentioned?

Solution: Note that it doesn't matter what the temperature or tension was. Whatever the values were, it is assumed that corrections for systematic errors were made. The computations here are for the random errors resulting from these sources.

$$e_{L_t} = \pm K_t L e_t \tag{5.11}$$

where

$$K_t = 0.00000645 \text{ ft/ft/°F}$$

$$L = 1000. \text{ feet}$$

$$e_t = \pm 5\text{F°}$$

from which

$$e_{L_t} = \pm 0.032 \text{ ft.}$$

$$e_{L_p} = \pm \left(\frac{\ell}{AE_m}\right) e_p \sqrt{n} \tag{5.13}$$

where

$$\ell = 100 \text{ ft.}$$

$$AE_m = 174{,}000 \text{ lbs (average values for A\&E}_m \text{ assumed)}$$

$$e_p = \pm 2 \text{ lbs.}$$

$$n = 10$$

from which

$$e_{L_p} = \pm 0.004 \text{ ft.}$$

The total uncertainty is

$$e_L = \sqrt{e_{L_r}{}^2 + e_{L_t}{}^2 + e_{L_p}{}^2} = \sqrt{0.045^2 + 0.032^2 + 0.004^2}$$

$$= \pm 0.055 \text{ ft.}$$

Students should form the habit of analyzing error situations to observe which sources are producing the most error in final results. This type of mental exercise will be helpful when Chapter 9 is studied. In this example it should be noticed that the major contribution in the total error was from reading and marking and the effect from tension was negligible. The next example will illustrate, however, that tension errors of ± 2 lbs. do not always result in negligible distance errors.

Example 5-9. Given: The same tape, total distance and error of Example 5-8, but with the situation changed in that end supported taping was employed because of rugged terrain. Assume that the tape operators can still mark the tape ends ± 0.01 ft. using plumb bobs. A tension of 15 lbs. was used at all times.

Solution: The only additional consideration is sag.

$$e_{L_s} = \pm \frac{w^2 \ell}{12P^3} e_p \sqrt{n} \tag{5.20}$$

where

$$w = 2 \text{ lbs. (assume average wt. of a tape)}$$

$$P = 15 \text{ lbs.}$$

$$e_p = \pm 2 \text{ lbs.}$$

$$\ell = 100 \text{ ft.}$$

$$n = 10$$

from which

$$e_{L_s} = \pm 0.062 \text{ ft.}$$

$$e_L = \sqrt{e_{L_r}^{\,2} + e_{L_t}^{\,2} + e_{L_p}^{\,2} + e_{L_s}^{\,2}} = \pm\ 0.083 \text{ ft.}$$

Note that the uncertainty from the sag source contributed the most to the total error. To reduce the effect, e_p could be reduced or P increased. It is far better to increase P rather than attempt too much care in reading tension. It can be shown that if P = 20 lbs. in this example, $e_{L_s} = \pm\ 0.026$ ft. and, at P = 25 lb., $e_{L_s} = \pm\ 0.013$ ft. which is approaching a value about which there shouldn't be much concern.

This example illustrates the point that taping skill is more than knowing how to use the tape and plumb bob. Precision is affected greatly by understanding the errors from a theoretical standpoint.

It should be emphasized that examples such as the above are for the purpose of illustrating error analysis, and the assumptions and error estimates won't fit all situations, nor are they necessarily complete for the particular exercises illustrated. For example, the assumption that tape operators can mark the tape ends ± 0.01 ft. using plumb bobs assumes much skill and ideal conditions. Even more important are the error sources possibly neglected. For example, the answer to Example 5-8 would have changed had an uncertainty for the tape's cross-sectional area been assumed. The results for Example 5-9 would have been different had an uncertainty for the tape's weight been assumed. The student may wish to investigate the effects of these variables on tape length.

As introduced in Section 4-6, errors can be propagated without the use of calculus. Using this alternate procedure, the uncertainty in the length due to sag in Example 5-9 can be calculated as follows:

$$C_s = \frac{2^2 \times 100}{24 \times 15^2} = 0.074 \qquad C_s = 0.074$$

$$C_{s+} = \frac{2^2 \times 100}{24 \times 17^2} = \underline{0.058} \qquad C_{s-} = \frac{2^2 \times 100}{24 \times 13^2} = \underline{0.099}$$

$$C_s - C_{s+} = 0.016 \qquad C_s - C_{s-} = 0.025$$

$$\text{Ave } e_{s_1} = \frac{0.016 + 0.025}{2} = 0.020 \text{ ft. (1 Tape length)}$$

$$e_{L_s} = 0.020\sqrt{10} = \pm\ 0.065 \text{ ft.} \quad \text{(10 tape lengths)}$$

It is noted that this closely agrees with the answer solved by the error equation in Example 5-9. It is also noted, that in order to get this answer, the correction for sag had to be calculated at 15 + 2 lbs tension and also at 15-2 lbs tension and the average of the differences between these corrections and 0.074 used. This is because the ± 2 lbs is appreciably large in comparison to 15 lbs. and the tension is squared rather than being only to the first power. It can be appreciated that the error equation for e_{L_s} is more straightforward and such derivations undoubtedly save time when the analyst has the calculus background. Naturally, once derived, an equation such as 5.20 can be used for any similar situation, employing only algebra, the calculus being used only for the initial derivation.

5-2 ANALYSIS OF STADIA MEASUREMENTS

Stadia Theory

Stadia was briefly discussed in Section 3-4 where it was explained that stadia measurements are made using the telescope of a theodolite which is equipped with stadia cross-hairs and a stadia rod or ordinary level rod.

The stadia principle is one of similar triangles. In Fig. 5.3,

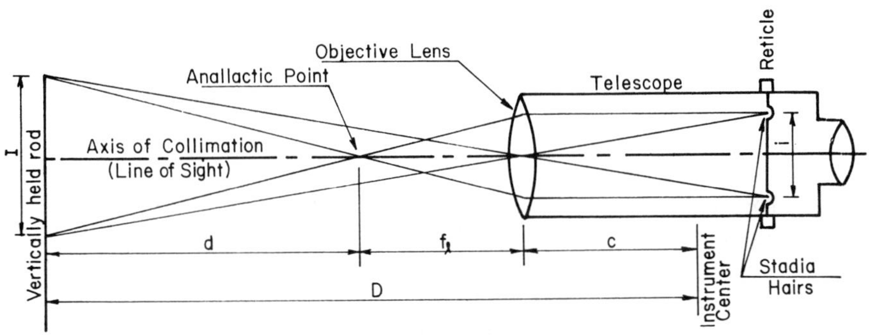

Figure 5.3 Stadia Principle

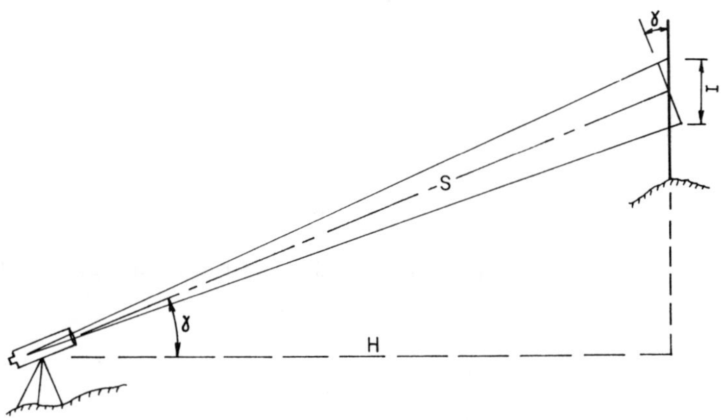

Figure 5.4 Stadia Distance with Inclined Sight

$$\frac{d}{I} = \frac{f_\ell}{i} \quad \text{from which} \quad d = I\,\frac{f_\ell}{i}\ .$$

Here, I is the rod intercept or difference between the upper and lower stadia hair readings, f_ℓ is the focal length of the telescope, and i is the distance betwen stadia hairs. The ratio f_ℓ/i is defined as the *stadia interval factor,* K. From Figure 5.3

$$D = d + f_\ell + c.$$

The sum ($f_\ell + c$) is defined as the *stadia constant*, C_{sc}. From these definitions and the two equations above,

$$D = KI + C_{sc} \tag{5.22}$$

The anallactic telescope was developed in 1823. It has, in effect a focal length equal and opposite to the instrument constant, c. Therefore, $C_{sc} = 0$. On modern instruments, telescopes are anallactic. They are commonly referred to as *internal focussing*. Older transit theodolite telescopes may have an *external focussing* telescope. For such a telescope, C_{sc} is usually assumed to be equal to one foot.

If terrain is fairly flat, stadia measurements can be made with the telescope horizontal. If terrain is sloping such that the lower cross-hair sighting is below the rod or the upper cross-hair appears over the top of the rod, the telescope must be raised or lowered so that both cross-hairs strike the rod in order to read the intercept. Thus, the telescope is no longer horizontal and the intercept is affected from what it would be if horizontal. When *inclined stadia readings* are made, the vertical or zenith angle must also be observed. It doesn't matter where on the rod the intercept appears for horizontal distance measurements. Figure 5.4 shows the situation for an inclined stadia measurement. From the geometry shown in the figure and Equ. 5.22,

$$S = KI \cos \gamma + C_{sc}$$

where

S = slope distance from the center of the telescope to the center cross-hair reading on the rod, and

γ = the vertical angle to the center hair reading.

Also,

$$H = S \cos \gamma$$

where

H = the horizontal distance from the instrument to the rod station.

From these two equations,

$$H = KI \cos^2 \gamma + C_{sc} \cos \gamma \tag{5.23}$$

If the common internal focussing telescope is used,

$$H = KI \cos^2 \gamma \; . \tag{5.24}$$

It is noted that other equations can be derived using zenith angle instead of vertical angle.

Systematic Errors

Corrections must be made to stadia measurements if the stadia rod has a systematic error or if K or C_{sc} are not their nominal values. Calibration of the rod, the telescope, and the stadia hair interval will reveal these errors.

A rod can be checked simply by measuring between its marks using a tape known to be accurate. The true values for K and C_{sc} can be discovered by establishing two accurate, but significantly different distances both of which are within range of the telescope for reading a precise intercept, then observing the intercepts over these distances. Two equations with unknowns K and C_{sc} are thus formed from which the values can be determined. Usually, however, minor deviations from nominal are unimportant since intercepts are normally no better than $\pm$ 0.01 ft. and thus if K $=$ 100, distances are accurate only to approximately the nearest foot anyway. For the purposes which stadia is employed, systematic errors are ordinarily neglected.

Random Errors

Although there is certainly an uncertainty in determination of K and C_{sc}, the random error in these values would ordinarily be quite negligible when accuracy of stadia is considered. There is also an uncertainty in centering the instrument and rod at their respective stations, but these errors are also negligible. Thus, the important variables are only I and γ.

Example 5-10. Given: A stadia distance is observed over sloping ground. The upper hair reading is 6.12 ft. and the lower hair reading is 3.41 ft. The vertical angle observed when these readings were taken was $-5^\circ\ 16'$. The random error in each hair reading is assumed to be $\pm$ 0.005 ft. and in the vertical angle, $\pm 30''$. What is the horizontal distance between the instrument and rod and the uncertainty in this distance?

Solution: Assume K $=$ 100 and $C_{sc} = 0$.

$$H = KI \cos^2 \gamma \tag{5.24}$$

where

$$I = 6.12 - 3.41 = 2.71 \text{ ft.}$$

$$\gamma = -5^\circ\ 16'$$

from which

$$H = 268.72 \text{ or } 269 \text{ ft.}$$

$$\frac{\partial H}{\partial I} = K \cos^2 \gamma$$

$$\frac{\partial H}{\partial \gamma} = 2KI \cos \gamma$$

$$e_H = \sqrt{(K \cos^2\gamma)^2 e_I^2 + (2KI \sin \gamma \cos \gamma)^2 e_\gamma^2}$$

where

$$e_I = \pm 0.005 \sqrt{2} \text{ (since intercept, } I = r_1 - r_2)$$

$$e_\gamma = \pm 30'' = 30/206{,}265 \text{ radians}$$

from which

$$e_H = \sqrt{0.4914 + 0.00005} = \pm 0.70 \text{ ft.}$$

From the magnitude of this error, it can be seen why stadia distances are expressed only to the nearest foot. The value might be given as 269 ± 0.7 ft.

From this example it can be seen that the error in the vertical angle has a negligible effect on the horizontal distance. The error in the distance is controlled primarily by the precision in the rod intercept. As can be expected, this precision is directly influenced by the distance, the distinctness of rod markings, natural conditions such as light and amount of haze, and the care taken in estimating readings. Precision can be increased slightly by setting the lower hair very carefully on an even mark (i.e., 2.00 ft., 3.00 ft., etc.) then estimating the upper reading. The uncertainty in the lower reading is probably less than in the upper. This technique also has the advantage that there is less likelihood of making an arithmetic mistake when subtracting the two readings.

Applications

Stadia measurements are useful in topographic surveying, in searching for lost control points, in referencing set monuments, for some photogrammetric control, and other applications where accuracy requirements are not particularly high. The primary advantage of the method is that it is fast.

Stadia can also be used to measure elevation differences. This will be covered in Chapter 7.

5-3 ANALYSIS OF SUBTENSE BAR MEASUREMENTS

Subtense Theory

The subtense bar, its field use and illustration was given in Chapter 3. Following is the theory and analysis of measurements with this instrument.

In Fig. 5.5, point A is the theodolite station and B is the remote station where the bar is centered, leveled, and set normal to the line of sight. From the geometry shown in the figure,

$$\frac{2H}{B} = \cot \frac{\alpha}{2}$$

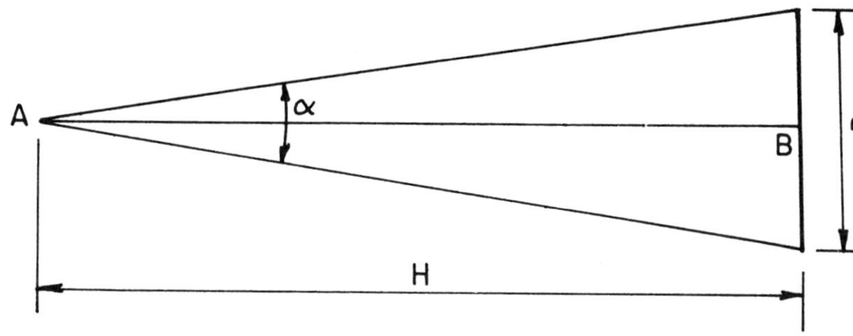

Figure 5.5 Subtense Principle

from which

$$H = \frac{B}{2} \cot \frac{\alpha}{2} \tag{5.25}$$

where

B = the distance between the targets on the bar,

α = the subtended angle between the targets,

H = the horizontal distance.

With a bar of 2 meters length, the formula is reduced to

$$H = \cot \frac{\alpha}{2} \text{ meters} \tag{5.26}$$

The value computed from these equations is the horizontal distance between the theodolite and bar stations regardless of the elevation difference between these stations since the measured angle is the horizontal angle. A common misunderstanding is that the horizontal angle changes with slope. It should be realized and remembered that, by definition, horizontal angle is the angle between two vertical planes. Therefore it doesn't matter where the instruments are located in the vertical line passing through the plane of pointing. The fact that there is no need for slope corrections is an important advantage of the subtense bar.

Systematic Errors

If there was an instrumental error in the bar length, this would cause an error in the length proportional to cot $\alpha/2$. Such an error could be due to either manufacture, damage to the instrument, or temperature effects on the bar length. Unless the bar had been abused it is doubtful that, due to care in manufacture and the low coefficient of thermal expansion, any appreciable systematic errors occur in subtense bar measurements. Therefore, most of the errors in subtense bar measurements are random in nature.

Random Errors

The uncertainties in distances determined by the subtense method are caused by errors in the subtended angle, the length of the bar, the bar's perpendicularity to the line being measured, and centering of both instruments. Equations to evaluate these effects will be derived in the following paragraphs.

Subtended angle error. Applying error propagation theory to Equ. 5.25 will yield an error equation in terms of the subtended angle error.

$$H = \frac{B}{2} \cot \frac{\alpha}{2} \tag{5.25}$$

$$\frac{\partial H}{\partial \alpha} = \frac{B}{2}\left(-\csc^2 \frac{\alpha}{2} \cdot \frac{1}{2}\right) = - \frac{B}{4 \sin^2 \alpha/2}.$$

Since $\sin \frac{\alpha}{2} \cong \frac{B}{2H}$ and α is small, the term

$$\frac{1}{\sin^2 \alpha/2} = \frac{4H^2}{B^2}.$$

By substitution,

$$\frac{\partial H}{\partial \alpha} = - \frac{H^2}{B}$$

From the general formula for propagation of random errors,

$$e_{H_\alpha} = \pm \frac{H^2}{B} e_\alpha \tag{5.27}$$

where

e_α = the error in the mean value of the angle, α, used to compute H

e_{H_α} = the error in the distance, H caused by error, e_α.

It should be realized that e_α is caused by many things, including reading, pointing, focussing, and various other natural, personal, and instrumental factors. Angular errors will be more thoroughly discussed in Chapter 6. For the examples to follow in this section, estimates of e_α will be made to demonstrate the use of Equ. 5.26. In this equation, the units of e_{H_α} are the same as the units of H and B, which must be the same. The units of e_α are radians.

Error in the bar length. No instrument is constructed with perfection. Although the user could not easily check or calibrate the subtense bar for removal of systematic errors, manufacturers generally give the possible error in a bar's length so that the quality of measurements can be analyzed. The subtense bar made by Wild-Heerbrugg has been advertised to be 2 meters in length to a precision of ±0.1 mm., a value too small to discover as to whether it is plus or minus or exactly how large.

$$\frac{\partial H}{\partial B} = \frac{1}{2} \cot \frac{\alpha}{2},$$

but,

$$\cot \frac{\alpha}{2} = \frac{2H}{B}$$

Therefore,

$$\frac{\partial H}{\partial B} = \frac{H}{B}$$

From the general formula for propagation of random errors,

$$e_{H_B} = \pm \frac{H}{B} e_B \tag{5.28}$$

where

e_B = the error in the bar's length,

e_{H_B} = the error in the distance, H caused by e_B.

The units of e_{H_B} are the same as the units of e_B. H and B must have the same units. If $e_B = \pm 0.1$ mm and $B = 2$ meters,

$$e_{H_B} = \pm 0.00005 \text{ H}. \quad (5.29)$$

The units of e_{H_B} are the same as the units of H in this equation.

Centering the instruments. If either the theodolite or subtense bar are centered over a point other than the true one, an erroneous distance will be determined. The centering error (commonly called *eccentricity*) is random in nature but it is not compensatory since the instruments remain in place throughout a series of measurements. The maximum error would occur if the eccentricity of both instruments happened to occur along the line being measured and in opposite directions. In actuality, the eccentricity at each station could occur at any point around the true station on a circle of radius equal to the error. A simple analysis would indicate that the maximum effect would rarely occur. Effects of eccentricity on angle measurement will be covered in Chapter 6. The explanations to be presented there can be applied to subtense bar measurements.

Bar alignment error. Analysis of this error reveals a behavior not conveniently classified as either random or systematic. Surely, the alignment itself, using the small scope sight is a random process since the operator attempts to point directly toward the theodolite to cause the bar to be perpendicular to the line. But, as can be seen from Fig. 5.6, up to an angular alignment error of $\alpha/2$, the pointings to the two ends of the bar shift a small and nearly equal amount either to the right or left depending on whether the bar was clockwise or counterclockwise in misalignment. Thus, although the pointings shift, they do so in the same direction and therefore there is no noticeable change in the subtended angle. But, as the alignment error exceeds $\alpha/2$, the pointings begin to shift in opposite directions, and always in the direction to decrease the subtended angle, thus causing a distance longer than true.

Experience has shown that the personal pointing error of the subtense bar toward the theodolite, assuming the scope sight is in adjustment, should not exceed about 10 cm. Since the pointing error would need to be 1 meter in order to have an angular alignment error of $\alpha/2$, the error should remain well within the range where the effects cancel. Thus, it can be concluded that errors in distance from this source need not be considered further.

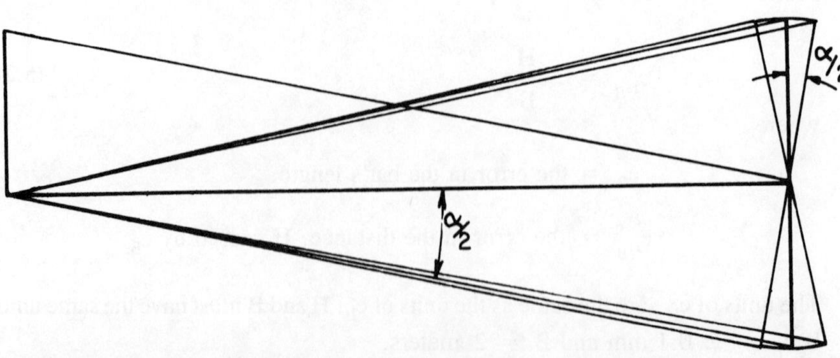

Figure 5.6 Subtense Bar Alignment Error

Example 5-11. Given: Several repetitions of a subtended angle between the targets of a 2-meter subtense bar yielded a mean angle of 1° 32′ 06″ with an uncertainty in the mean of ± 1.5″. The manufacturer's error in the length of the bar was stated to be ± 0.1 mm. Ignoring eccentricity and other errors, what is the length of the line and the uncertainty in the length?

Solution:

$$H = \frac{B}{2} \cot \frac{\alpha}{2} \tag{5.25}$$

where

$$\frac{B}{2} = 1 \text{ meter,}$$

$$\frac{\alpha}{2} = 0° \; 46' \; 03''$$

from which

$$H = 74.648 \text{ m.}$$

$$e_{H_\alpha} = \pm \frac{H^2}{B} e_\alpha \tag{5.27}$$

where

$$H = 74.65 \text{ m}$$

$$B = 2 \text{ m.}$$

$$e_\alpha = \pm 1.5'' = \pm 0.000007272 \text{ radians}$$

from which

$$e_{H_\alpha} = \pm 0.020 \text{ m.}$$

$$e_{H_B} = \pm 0.00005 \text{ H} \tag{5.29}$$

where

$$H = 74.648 \text{ m}$$

from which

$$e_{H_B} = \pm\ 0.004 \text{ m}$$

$$e_H = \sqrt{e_{H_\alpha}{}^2 + e_{H_B}{}^2} = \pm\ 0.020 \text{ m.}$$

Thus, the distance could properly be stated as 74.65 ± 0.02 meters. It is noted that the error due to bar length contributed very little to the total error in this example.

Applications

The subtense bar is suitable for measurements across inaccessible or rough terrain (rivers, busy streets, etc.). Its precision, as the homework exercises at the end of the chapter should indicate, is high for fairly short distances if a precise theodolite is used. The advantage of having the horizontal distance directly without need to consider slope is an additional advantage.

5-4 ANALYSIS OF ELECTRONIC DISTANCE MEASUREMENTS

Nature of Electronic Distance Measurements

Measurement of distances electronically involves physical laws fundamentally different than the mathematical and physical principles employed for nearly all other measurement methods discussed in this book. The natural physical factors such as tension, temperature, sag, refraction, slope and earth's curvature which affect most surveying measurements are fairly easily understood and quantifiable. Likewise, the mathematics of direct and indirect measurement is mostly simple arithmetic and trigonometry. However, electronic distance measurement (EDM), if fully understood by the user of the equipment, requires a background in electronics and physics beyond that needed for most other surveying measurement. But, unlike other modern developments such as photogrammetry, for example, surveyors can employ EDM without understanding very much of the physical phenomena involved. This is true in part because EDM equipment has become very automatic and also in part because the accuracy, relatively speaking, is much less at the mercy of the three common sources of error. Instrumental failure normally results in no readings at all, or data so erroneous that malfunction is obvious. Although the principle of operation is more complicated than with traditional instruments, the accurate use of the equipment requires less operator skill. For this reason, it makes little sense to present, in a text such as this, the electronics or the principle of operation beyond what has already been presented in Chapter 3. The errors in EDM which can be checked and corrected by the surveyor are adequately covered in references on EDM listed at the end of Chapter 3. Few of the errors lend themselves to the type of analyses being developed herein. They will be briefly discussed, however.

Instrument Constant

This error occurs if the electrical center of the instrument is not along the same vertical line as the physical center of gravity which is plumbed over the station. The value of this error is furnished with the equipment but is subject to change.

Discovering the constant is done through calibration. There are two calibration techniques. These techniques are adequately explained in references on EDM at the end of Chapter 3 and will not be covered here.

Once the instrument constant is determined through calibration, a correction is applied to all measurements. Relative to the distances normally measured, the error in the constant is usually negligible and would therefore not represent an uncertainty of any concern.

Reflector Constant

For light wave instruments, there is an error in the distance caused by the fact that the plumb line through the reflector's physical center of gravity is different from the optical center of the prism. Reflector constants are furnished with the equipment, but can be checked as explained in the references.

Corrections for this error are applied to measurements in a manner similar to corrections for the instrument constant. Uncertainties in the constant should be quite negligible.

Manufacturer's Error

This is an uncertainty factor published in the manuals and brochures for each model of EDM. Values vary between models. Usually, the error is stated as a constant plus or minus value plus an additional uncertainty which is relative to the distance. For example:

$\pm\left(0.02 \text{ ft.} + \frac{D}{500{,}000}\right)$ where D is the distance in feet. This uncertainty represents one of the principal random errors in EDM. Generally speaking, the more expensive the instrument, the lower the error. For very short distances, it can be seen that some EDM may not be as precise as other methods, such as taping or subtense bar, because of the constant uncertainty.

Natural Errors

Temperature and atmospheric pressure affect the speed of light and microwaves, thus causing an error. Microwaves are also affected by humidity. Thus, the user must read these variables when measuring and account for their effects. They are corrected either by using nomograms which give the errors for each source as a function of distance and the variable itself, or by setting dials on the EDM equipment to the values read for each natural error source. The latter method automatically corrects the reading prior to read-out and is common on modern instruments. Uncertainties from these natural sources are difficult to evaluate, but would normally be negligible if properly corrected.

All EDM measures slope distance. Therefore, another natural error is slope, the final desired value generally being horizontal distance. This is normally corrected using trigonometry as with slope taping. The vertical angle is measured at one or both ends of the line, which, along with the measured slope distance, gives data for the correction. The horizontal distance can also be determined from station elevations if known, using the same equation as in taping, when the vertical distance between tape ends is known. Uncertainties in horizontal distance are caused by random errors in the slope distance and the vertical angle or elevation difference.

Personal Errors

About the only personal errors in EDM are those associated with slope corrections and eccentricity of the instruments over the stations. The latter are negligible when the range of EDM is considered.

Example 5-12. Given: A light wave type EDM with direct read-out is used to measure a distance. The reading of the slope distance is 2,462.815 ft. The EDM is centered over one station at a height of 4.4 ft. The reflector system is centered over the remote station at a height of 5.1 ft. A zenith angle of 90° 49′ 16″ is read with a theodolite at the EDM station. The theodolite has an h.i. of 5.2 ft. The angle is measured by pointing to the center of the reflector. The manufacturer lists an error of $\pm\,(0.01 \text{ ft.} + \dfrac{D}{500{,}000})$ where D is the slope distance. Atmospheric corrections are compensated automatically by the equipment. The instrument constant is + 0.015 ft. and the reflector constant is − 0.023 ft. What is the horizontal distance and the uncertainty in the distance?

Solution: The systematic errors are the instrument constant, the reflector constant, and the slope.

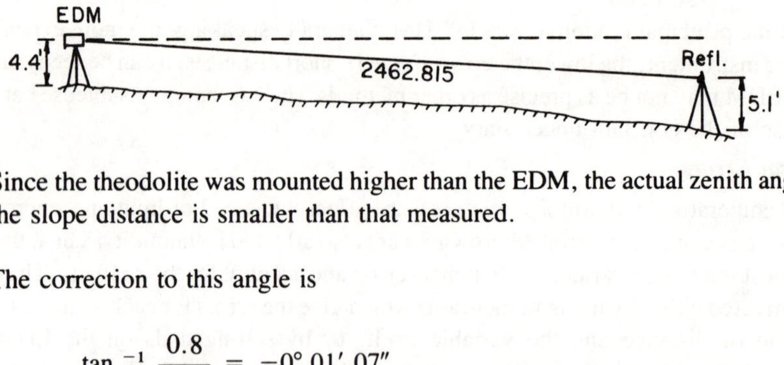

Since the theodolite was mounted higher than the EDM, the actual zenith angle of the slope distance is smaller than that measured.

The correction to this angle is

$$\tan^{-1} \frac{0.8}{2463} = -0°\,01'\,07''$$

$$z = (90°\,49'\,16'') - (0°\,01'\,07'') = 90°\,48'\,09''$$

$$H = S \sin z$$

where

$$S = 2462.815$$

from which

$$H = 2462.573 \text{ ft.}$$

The true distance between ground stations must be corrected for the instrument and reflector constants.

$$H_T = 2462.573 + 0.015 - 0.023 = 2462.565 \text{ ft.}$$

The uncertainties are caused from the manufacturer's error and the zenith angle measurement.

$$e_m = \pm (0.01 \text{ ft} + \frac{2463}{500{,}000}) = \pm 0.015 \text{ ft.}$$

$$e_g = \pm S \cos z \, e_z$$

Assuming an angle error of ±1 second yields

$$e_g = \pm 0.0002 \text{ (negligible)}$$

Therefore, the uncertainty in the distance is approximately ±0.015, assuming no other error sources. This yields a relative error of about 1:165,000. It can be seen why there need not be much concern about uncertainty in EDM measurements.

Problems - Chapter 5

Work Problem 5.1 to 5.6 based on the following:

A 50 meter steel tape has been calibrated at 20°C and 5 Kg tension, fully supported. The calibrated length is 49.997 m. The tape's weight is 1.5 Kg. Use average values of modulus of elasticity and cross-sectional area given in Chapter 3.

5.1 What will be the length of this tape at 30°C with the same tension and support condition as used in calibration?

5.2 What will be the horizontal distance between the end points of this tape at 30°C, end supported, 5 Kg tension?

5.3 What will be the horizontal distance between the end points of this tape at 30°C, 7 Kg tension, and supported at the 0, 25, and 50 m marks?

5.4 A distance of 268.20 meters is to be established using this tape. The field temperature is −4°C. Using the tape fully supported, what should be the final reading to establish this distance?

5.5 Assuming an uncertainty in reading and marking at each tape end of ± 2 mm., an uncertainty in tension of ± 0.25 Kg., and uncertainty in field temperature of ± 2°C, what is the uncertainty in the distance established in Problem 5-4?

5.6 Assuming an uncertainty of ± 0.25 Kg in determining tension, what would be the uncertainty in a distance of 450 meters if the tape is used as described in Problem 5.2?

Work Problems 5.7 through 5.10 based on following:

A 100 ft. steel tape is known to be 100.003 ft. long at 70°F, 15 lbs tension, fully supported. Use average values for modulus of elasticity and cross-sectional area given in Chapter 3.

5.7 If this tape was calibrated, end supported at 70°F, 15 lbs tension and found to be 99.956 ft. between the 0 and 100 ft. marks, what is its weight?

5.8 What tension should be applied to this tape in order to counteract systematic errors if the field temperature is 52°F and the tape is to be used fully supported?

5.9 A distance of 865.35 feet was observed between two existing points with this tape, using it fully supported at 80°F. The percent slope for each segment of the distance was observed to be 0% for the first 325 feet, 2.8% from 325 to 400, 3.5% from 400 to 500, – 1.8% from 500 to 600 and –4.2% from 600 to end. These observations were made using a hand-held surveyor's clinometer at each change in slope. What is the true horizontal distance between the two points?

5.10 In problem 5.9 assume a reading and marking uncertainty of ± 0.005 ft. per tape end, a field temperature uncertainty of ± 3°F, a tension uncertainty of ± 2 lbs, and each segment of slope (there are 5 segments mentioned) uncertainty of ± 1%. Compute the uncertainty in the horizontal distance.

For the stadia Problems 5.11 and 5.12, use K = 100 and $C_{sc} = 0$.

5.11 Assuming an instrument operator can place the lower stadia hair reading on a rod mark to a precision of ± 0.002 ft. and then read the upper mark to ± 0.005 ft., what is the uncertainty in a stadia distance of about 300 feet?

5.12 A metric rod was used for stadia observations. The upper hair reading was 3.12 meters and the lower hair reading was 1.82 meters. The vertical angle was 0° 42′ . Compute the horizontal distance.

5.13 Plot a graph of errors in subtense bar distance measurements using values of e_α = ± 1.0″, 2.0″, and 5.0″, and error in the bar length of ± 0.1 mm. Plot errors in distance on the x axis and distance on the y axis. (Note: There will be three lines — one for each value of e_α.)

5.14 Assuming an uncertainty in subtended angle of ± 1 second, what would have been the uncertainty in the length in Problem 5.4 using the subtense bar method? Compare this answer with that of Problem 5.5 and with what would likely have been the result using typical EDM.

Additional Reading References - Chapter 5

1. Barry, Austin B. *Errors in Practical Measurement in Science, Engineering, and Technology.* New York: John Wiley & Sons, Inc., 1978, Chapter 7.

2. Brinker, Russell C. and Wolf, Paul R. *Elementary Surveying*. Sixth Ed. New York: Harper and Row, 1977, Chapter 4.

3. Buckner, R.B. "Survey Distance Measurements and When to Use What", *Proceedings of the American Congress on Surveying and Mapping,* Falls Church, Virginia, March 1970, pp. 1-19.

4. Moffit, Francis H. and Bouchard, Harry. *Surveying*. Seventh Ed. New York: Harper and Row, 1981, Chapter 2.

5. Schmidt, Milton O. and Rayner, William H. *Fundamentals of Surveying*. Second Ed. New York: D. Van Nostrand Co., 1978, Chapter 2.

CHAPTER 6

ANALYSIS OF ANGLE MEASUREMENTS

6-1 UNDERSTANDING THE THEODOLITE

General Concerns

The entire discussion of angle measurement will focus on analysis of horizontal angle measurements using repeating and direction optical theodolites. Much of the discussion of repetition instruments also applies to vernier theodolites (transits) except for the analysis of reading errors. Instrument classification, basic measurement procedures, sample field notes, and other pertinent definitions were covered in Section 3-3. The meaning of *horizontal angle, direction, azimuth,* and *bearing* were given in Section 1-6. Beginning students may wish to review those sections before proceeding further.

Adjustments of vernier thoedolites have traditionally been covered in surveying textbooks since the mid-nineteenth century. Optical theodolites stay in adjustment much better than the traditional American transit. Most major adjustments of optical theodolites are checked and corrected at authorized service depots whenever an instrument is returned for repair, cleaning, or reconditioning. Thus, the user is not as concerned about adjustment of these instruments as with the more crude vernier instruments. This text will cover the user tests for maladustment so that an appreciation of the instrumental errors is gained. Most of the actual adjustments are best made by qualified instrument technicians. Those that can be made by the user are explained in manuals for each particular model of instrument. The error sources in horizontal angle measurements are primarily personal and instrumental. Most are random in nature, but a few are systematic. As usual, most of the random errors are from personal sources and the systematic errors are usually caused by instrument maladjustment. It will be seen that most of the important instrumental errors can be compensated by proper use of the instrument. Furthermore, modern theodolites are constructed with much care and refinement in finishing and fitting the parts, so that they do not become maladjusted except by extreme abuse. Nevertheless, the student should be aware of the geometry of construction of the theodolite in order to analyze measurements in a professional manner and to know the instrument itself for the sake of professional pride and artful application.

Basic Instrument Construction

Definitions. There are four principal lines or axes that must be in their proper relationship in a theodolite. The *line of sight* (also called *collimation axis* or *line of collimation)* is a line connecting the intersection of the cross-hairs and the optical center of the objective lens of the telescope. The *bubble axis* is the line tangent to the upper inside surface of the plate bubble tube at its midpoint. These terms were defined in Section 3-2 where the telescope and bubble tube were illustrated. The *vertical axis* of a theodolite is the axis of rotation or the centerline of the inner spindle. The *horizontal axis* (also called *transit axis* or *axis of trunnion)* is the axis of the trunnion supporting the telescope. It is this axis which adds the "transiting" or "plunging" feature to the theodolite. These instrument axes are all illustrated in Fig. 3.12, where other construction features are also shown. Details of instrument construction vary among models but these definitions and features illustrated are the same in all models.

Desired relationships.

1. The vertical axis must be perpendicular to the bubble axis so as to have the vertical axis truly vertical when the bubble is centered.
2. The line of sight must be perpendicular to the horizontal axis so that the line of sight revolves in a vertical plane rather than along a conical surface.
3. The horizontal axis must be perpendicular to the vertical axis so that the horizontal axis is parallel with the bubble axis and the line of sight passes through a vertical plane rather than an inclined one when the telescope is rotated about the horizontal axis.
4. The vertical cross-hair should be vertical and the horizontal cross-hair should be horizontal when the instrument is leveled.
5. The vertical axis, horizontal axis, and line of sight should intersect at one point.
6. The horizontal circle should be mounted perpendicular to the vertical axis.
7. The vertical axis should pass through the center of the horizontal circle.
8. Graduations of markings around the circles should be evenly spaced.
9. The optical plummet axis should be coincident with the vertical axis and perpendicular to the plate bubble axis.

The first three of these are the most important and can all be tested by the user.

6-2 INSTRUMENTAL ERRORS IN ANGLE MEASUREMENTS

Bubble Adjustment

Of the several instrument maladjustments, this is the one that is most readily seen by the observer. If the bubble axis is not perpendicular to the vertical axis, the bubble will move when the instrument is pointed in different directions. If not in adjustment, recentering it *does not* bring the vertical axis into vertical and thus the horizontal circle is not rotating in a horizontal plane.

With the instrument mounted on the tripod and the bubble centered in one position, the vertical axis might appear as in Fig. 6.1 (a) if it was not perpendicular to the bubble axis. This would be discovered when the instrument was rotated about the vertical axis. If it was rotated exactly 180° from the position shown in Fig. 6. 1 (a), the relationship between the bubble axis and vertical axis would be as shown in Fig. 6. 1 (b). Note that, in this position, the vertical axis remains in its same non-vertical alignment and the bubble would now be off center. The mala-dustment is exaggerated in the sketch. The bubble marks provide a means to detect the non-perpendicularity and the amount, in units of bubble divisions. It is noted that, as between position (a) and (b) in this figure, the bubble moved through an angle $2E_b$. If the number of divisions it moved is counted, it can be moved back half that amount using the leveling screws, thus rotating it *and* the vertical axis through an angle, E_b, which makes the vertical axis vertical. Then, for all angular measurements, if the bubble is kept in this same off-center position with the leveling screws when pointing the telescope in any direction, the vertical axis will remain vertical and thus, the horizontal circle should be horizontal. Actual adjust-ment of the bubble tube is a better procedure, however. After observing the amount the bubble moves in Fig. 6.1 (b), the appropriate capstan screws can be manipulated, following instructions in instrument manuals, to move the bubble axis through the angle E_b with respect to the vertical axis. This would make the two axes perpendicular. Then, the bubble is leveled and kept centered using the leveling screws and the vertical axis thereby kept vertical throughout the measur-ing procedures.

The effect of this maladjustment is difficult to quantify. Ignoring non-verti-cality of the vertical axis when measuring horizontal angles causes an error which is proportional to the inclination of the line of sight (vertical angle), the size of the angle, the amount of maladjustment, and the bubble sensitivity. The error is not compensated by averaging direct and inverted readings. The vertical axis *must* be kept vertical throughout the angle measuring procedure in order for the circle to remain horizontal. If procedures as described above are employed to correct or compensate for this instrumental error, its effects should be negligible for most work. More important is the related personal error of centering the bubble, which will be discussed later.

Line of Sight Error

This maladjustment is not readily detected while operating the instrument, but a test can be made to discover it. Refer to Fig. 6.2 for the test. With the instrument mounted on the tripod at point I, and plate bubble centered the instrument is sighted to a distant point A, then the telescope plunged and pointed in the opposite direction and a distant point B is set on the line of sight. If the line of sight is not perpendicular to the horizontal axis, point B will not be on a straight line from A through the instrument station. The amount of linear error depends on the amount of maladjustment and the distances involved. Unless the instrument has been damaged, BC should be very small–perhaps only a few centimeters. Keeping the telescope in the inverted position, point A is again sighted by rotating the instru-ment about the vertical axis. After sighting on point A, the telescope is plunged

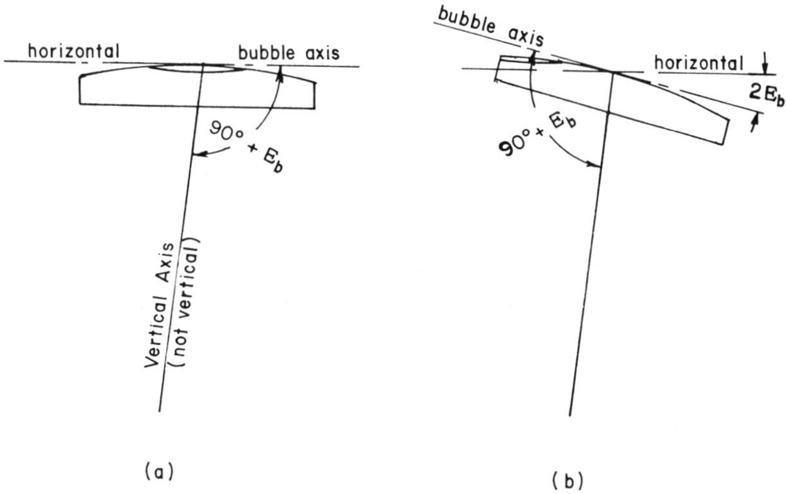

Figure 6.1 Theodolite Plate Bubble Adjustment

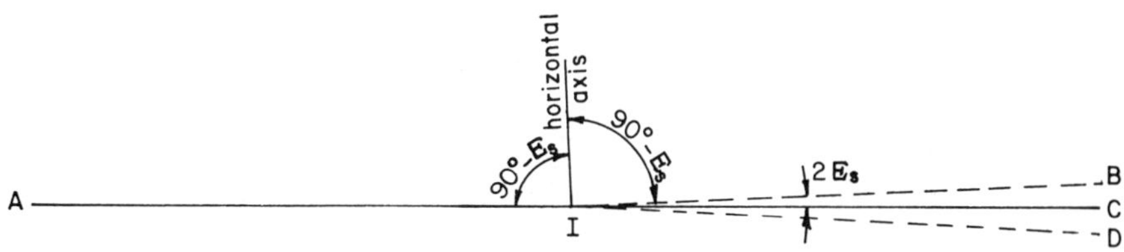

Figure 6.2 Theodolite Line of Sight Adjustment

and a point D is aligned near B.

Point C represents a point on the line AI extended. The error angle $\mathbf{E_s}$ is the maladjustment between the line of sight and the horizontal axis. The position of point C is unknown until the test is completed. When sighting D, the telescope is erect instead of inverted as when B was aligned, so the error angle is reversed. Thus, CD should be equal to BC.

If a true straight line was being established, it would be a simple matter to set point C midway between B and D. To adjust the instrument so that points B and D coincide at C, the line of sight (vertical cross-hair) must be moved. Capstan screws are adjusted following specific instructions in the instrument manual. It can be seen that, in order to correct the error angle, $\mathbf{E_s}$, the line of sight must be moved through ¼ the distance from D toward B, since BD subtends an angle 4$\mathbf{E_s}$. The angle of maladjustment, $\mathbf{E_s}$, can easily be determined by measuring CI and BD.

In the above test, the instrument and stations A and B should be at approximately the same elevation. The distances AI and IB should be at least 100 meters.

The effect of maladjustment of the line of sight is that it traces an imaginary conical surface when the telescope is rotated about the horizontal axis. The base of this imaginary cone is vertical and at infinity. Its axis is the horizontal axis of the theodolite. It is a very short cone with the angle at the apex equal to $180° - 2E_s$. Although the adjustment error, E_s, remains constant when an angle is measured, the actual horizontal angular error in a given pointing varies with the angle of inclination since the line of sight is revolving in a conical pattern.

Figure 6.3 illustrates the geometry of the variation in horizontal angular error of a pointing with vertical angle variation. In this figure point 0 represents the intersection of the vertical and horizontal axes of a theodolite. Line OA is a line perpendicular to the horizontal axis, or an imaginary line which would be the line of sight if it was adjusted properly. Line OB is the actual line of sight. Both OA and OB are horizontal. Thus E_s is the error angle between the line of sight and the horizontal axis, and represents the error in a direction when pointing along a horizontal plane. When the telescope is inclined (or depressed) through a vertical angle, γ, the line of sight along this inclined angle has an error in direction of E_c to target C. Line CD is vertical.

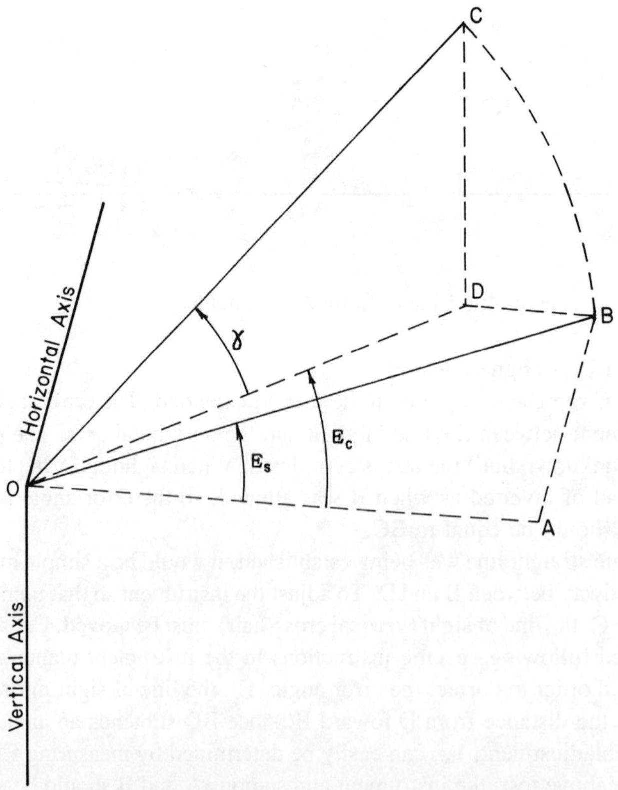

Figure 6.3 Geometry of Line of Sight Error

An analysis of the combined effects of inclination of the line of sight and maladjustment of the line of sight can be made as follows. Assume OB = OC.

$$\sin E_s = \frac{AB}{OB} = \frac{AB}{OB} \cdot \frac{OD}{OD} = \frac{AB}{OD} \cdot \frac{OD}{OB}$$

$$\sin E_s = \frac{AB}{OD} \cdot \frac{OD}{OC} = \sin E_c \cos \gamma$$

Or,
$$\sin E_c = \sin E_s \sec \gamma$$

Since E_c and E_s are small the ratio of their sines are nearly equal to their angular ratio. Therefore,

$$E_c = E_s \sec \gamma \tag{6.1}$$

where E_c = horizontal angular error in a pointing resulting from combined effects of the line of sight maladjustment E_s and vertical angle γ.

If a single horizontal angle was measured, it can be seen that the actual error in the angle would be

$$E_{\alpha_c} = E_{c_1} - E_{c_2}$$

where E_{c_1} and E_{c_2} are errors in pointings to directions 1 and 2. The true value of the angle would be $\alpha_D + E_{c_1} - E_{c_2}$ where α_D is the difference in readings to the backsight and foresight targets (1 and 2). If the vertical angles are the same in sign and magnitude, the errors would cancel. The special case of this is when the vertical angles are zero (flat terrain), in which case $\sec \gamma = 1$, $E_c = E_s$ and $E_{s_1} = E_{s_2}$. From the above, it can be deduced that measurements of single angles in hilly terrain will have larger errors than measurements in flat terrain. The larger the vertical angle difference between backsight and foresight, the larger the error in the horizontal angle due to any line of sight maladjustment.

Let us next analyze what occurs when the angle is doubled, measured once with the telescope direct and once with it inverted. The direction of the error is reversed when the telescope is inverted, the actual value of a single angle with the telescope inverted being

$$\alpha_R - E_{c_1} + E_{c_2}$$

where α_R is the difference in readings to the backsight and foresight stations. Thus

$$\alpha_T = \frac{(\alpha_D + E_{c_1} - E_{c_2}) + (\alpha_R - E_{c_1} + E_{c_2})}{2} = \frac{\alpha_D + \alpha_R}{2}$$

where α_T is the true value of the angle. It can be seen that any error due to maladjustment of the line of sight, regardless of vertical angle values or differences, is compensated by taking an equal number of readings with the telescope in the direct as in the inverted position and merely using the arithmetic mean of the sum of the readings as the true value. In the last equation, α_D and α_R would appear as individual values with a direction theodolite but would be an accumulated sum within the instrument using a repetition instrument.

Horizontal Axis Error

If the horizontal axis is not perpendicular to the vertical axis, the line of sight will pass through an inclined rather than a vertical plane when the telescope is rotated about the horizontal axis. This maladjustment can be discovered by a simple test. Refer to Fig. 6.4. The instrument is set on the tripod located about 20 to 50 meters from a building. The plate bubble must be centered and properly adjusted perpendicular to the vertical axis. Any well defined point A, located on the building at a vertical angle of 30° - 50° is first sighted. The telescope is then lowered to point horizontally toward the building. With a pencil or other device, point B is marked on the building. Point D is in a vertical plane through A and represents the point to which the telescope would point if the horizontal axis was perpendicular to the vertical axis. Point D's location is undefined at this stage. The next step is to invert the telescope and again point to A. Next, the telescope is lowered to a point at the same height as B. This point, C, will lie on a plane inclined with respect to the vertical an equal and opposite amount as that of the plane through AB. Thus, if point C is marked, the position of D is simply midway between B and C. The angle of inclination of the horizontal axis is the angle BAD or DAC.

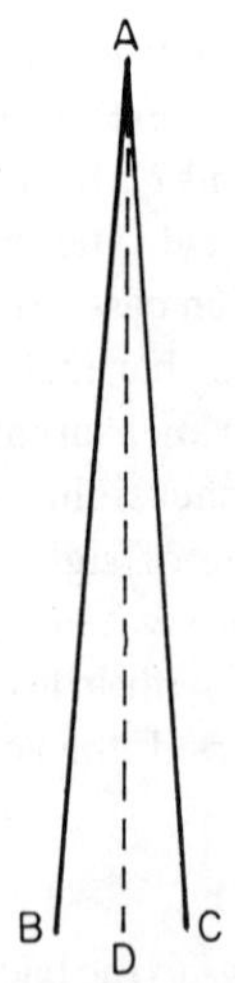

Figure 6.4 Theodolite Horizontal Axis Adjustment

Adjustment of the horizontal axis of an optical theodolite is normally not done by the user, although the above test can easily be performed to see if a maladjustment exists for adjustment by qualified technicians. Most American vernier instruments have capstan screws at one end of the horizontal axis for adjusting it. The adjustment is performed by first pointing to D using the fine motion screw of one of the vertical axis motions. Then, the telescope is raised to the elevation of point A which will move it through an inclined plane through D and parallel with

CA. The axis is adjusted by moving the appropriate screws until the line of sight rotates through the angle of maladjustment and points to A. After adjustment, the telescope should pass through the vertical plane AD when it is raised or lowered. It is important to keep the plate bubbles centered throughout the test.

The effect of maladjustment of the horizontal axis is that the line of sight doesn't pass through a vertical plane, which it must do if true horizontal angles are to result since they are, by definition, angles between vertical planes.

Figure 6.5 illustrates the horizontal angular error for one pointing caused by the combined effects of vertical angle and inclined horizontal axis. In this figure, the point of intersection, O, of the two planes is on a horizontal line through the telescope of the theodolite which should be imagined as located a horizontal distance, D, from that point, facing the figure. To point to target P, the line of sight is inclined downward through a vertical angle, γ. The maladjustment of the horizontal axis is angle E_h. The vertical distance through which the line of sight passes in order to point to the target is v. The actual horizontal displacement of the line of sight is x. That is, although the observer has centered the line of sight on the target, a fictitious target exists at a point located x units away along a vertical plane through the telescope. An analysis of the combined effects of inclination of the line of sight and maladjustment of the horizontal axis can be made as follows.

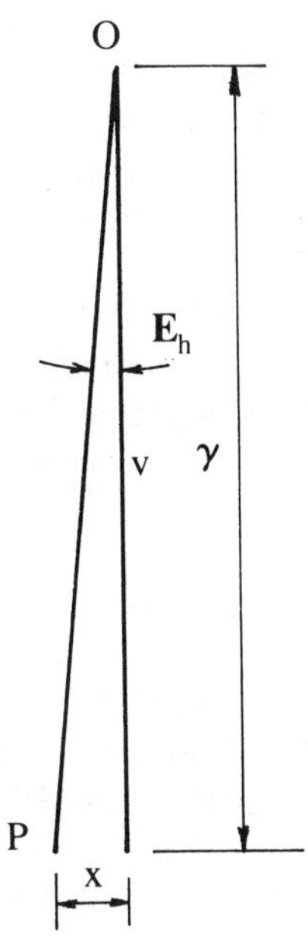

Figure 6.5 Geometry of Horizontal Axis Error

$$v = D \tan \gamma$$
$$x = E_h v = E_h D \tan \gamma$$
$$E_H = \frac{x}{D}$$

Therefore,

$$E_H = E_h \tan \gamma \tag{6.2}$$

where E_H = the horizontal angular error in a pointing resulting from combined effects of horizontal axis maladjustment E_h and vertical angle γ. It can be seen that the angular error in the pointing is independent of the distance, D, and is solely a function of the maladjustment of the horizontal axis and the vertical angle. The value E_h can be determined by the instrument test based on Figure 6.4. Distance, D, vertical angle, γ, and x (the same magnitude as BD in Figure 6.4) must be measured. Then E_h can be solved from the above equations. This solution is left as an exercise for the student.

If a single horizontal angle was measured, it can be seen that the actual error in the angle would be

$$E_{\alpha H} = E_{H_1} - E_{H_2}$$

where E_{H_1} and E_{H_2} are errors in pointings to directions 1 and 2. As with the line of sight maladustment, the errors cancel if vertical angles are the same. If vertical angles are zero, there are no errors in the pointings. The larger the vertical angle difference between backsight and foresight, the larger the error in the horizontal angle. Hilly terrain would cause larger errors than very gently rolling or flat terrain, for a given horizontal axis maladjustment.

Let us next analyze what occurs when the angle is doubled, measured once with the telescope direct and once with it inverted. Borrowing from the previous discussion of the line of sight errors, it should be obvious that the error in pointing to a particular target is reversed when the telescope is inverted. The manner in which the errors cancel between direct and inverted readings is illustrated in Fig. 6.6. In this figure,

α_T = true horizontal angle between targets 1 and 2,
α_D = angle read between targets 1 and 2, telescope direct,
α_R = angle read between targets 1 and 2, telescope inverted,
x_1, x_2 = linear pointing error to targets 1 and 2, respectively,
E_{H_1}, E_{H_2} = angular pointing error to targets 1 and 2, respectively,
γ_1, γ_2 = vertical angle to targets 1 and 2, respectively,
E_h = angle of inclination of horizontal axis.

It is seen that

$$\alpha_T = \alpha_D - E_{H_1} + E_{H_2} = \alpha_R + E_{H_1} - E_{H_2}$$

or

$$\alpha_T = \frac{\alpha_D - E_{H_1} + E_{H_2} + \alpha_R + E_{H_1} - E_{H_2}}{2} = \frac{\alpha_D + \alpha_R}{2}$$

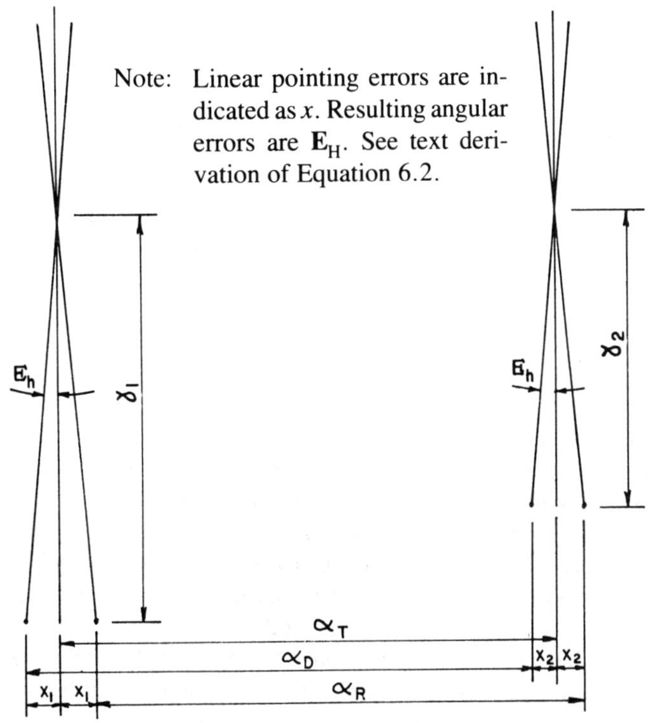

Figure 6.6 Compensation of Horizontal Axis Error

Thus, any error due to maladjustment of the horizontal axis, regardless of vertical angle values or differences, is compensated by taking an equal number of readings with the telescope in the direct as in the inverted position, and the true value of the angle is simply the arithmetic mean of the readings, or the accumulated reading divided by the number of repetitions.

Other Instrumental Errors

In section 6-1, six additional desired relationships were mentioned. Each of these will now be briefly discussed.

Cross-hairs. The verticality of the vertical cross-hair is of minor importance, depending on techniques of the observer. Even though it may be adjusted perpendicular to the horizontal axis at its point of intersection with the horizontal cross-hair, it might be rotated slightly. This would cause an error in a single pointing if other than the point of intersection of the cross-hairs was aligned on the target, as shown in Fig. 6.7.

However, if the same location along the vertical cross-hair was aligned on both backsight and foresight targets, the error would cancel for a single angle measurement even if the point was other than the cross-hair intersection point. For

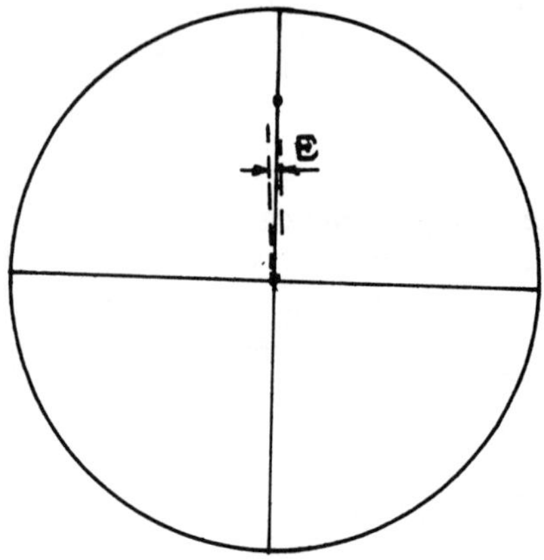

Figure 6.7 Cross-Hair Error Effect

a pointing with the telescope inverted, the cross-hair maladjustment would appear just as in the erect position. For example, if the reticle is rotated clockwise from its correct position, it will remain that way whether the telescope is erect or inverted. Taking sets of angles direct and inverted has no effect as regards compensating for this maladjustment. The only way to compensate for the error is to sight at the same point along the vertical cross-hair for the backsight and foresight of each single angle. Preferably, the point of intersection of the cross-hairs should be used as the sighting point. The cross-hair adjustment can be checked by leveling the plate bubbles, and with the bubble axis and horizontal axis in proper adjustment, sighting along a known vertical line (such as a plumb bob string) and observing the deviation of the vertical hair from this line. If provision is made on the instrument for adjusting the cross-hairs, the appropriate capstan screws can be loosened to do so, being careful not to translate the reticle which would disturb the line of sight adjustment. Instructions in manuals for each specific instrument should be closely followed.

Optical plummet axis. The optical plummet axis should be perpendicular to the plate bubble axis so that it points along a vertical line when the plate bubble is centered. An instrumental error having a systematic effect on a given set of angle measurements would result if this axis was maladjusted since the instrument would be off center with respect to the ground station. Since the construction of instruments varies, the method of testing this adjustment varies. It is a simple test and should follow specific instructions given in instrument manuals. A related personal instrument centering error should not be confused with this instrumental error, although their effects are the same on results. Instrument centering as a personal error will be discussed subsequently. Its analysis can be applied to the instrumental error effects.

Other adjustments. The remainder of the desired relationships listed in Section 6-1 are relatively unimportant. Unless the instrument has suffered from extreme abuse, the design, quality control, and construction of the instruments reasonably assure that the three main axes of the instrument intersect at one point, that the circle is mounted perpendicular to the vertical axis, that the vertical axis passes through the center of the horizontal circle, that the markings are evenly spaced around the circle, and that the optical plummet axis coincides with the vertical axis. Most of these maladjustments, if they exist, can be compensated either by the direct and inverted process or by observing at various locations around the circle for a series of measurements to be averaged for an angle.

6-3 PERSONAL ERRORS IN ANGLE MEASUREMENT

Focussing

Slightly out-of-focus cross-hairs cause an optical phenomenon known as *parallax*. Two situations cause parallax. One is when the eyepiece is focussed on the reticle plane but the objective is brought to focus in a plane slightly away from the reticle plane. The other situation is when the eyepiece and objective are focussed at a common point which is not in the reticle plane. The first situation is illustrated in Fig. 6.8 (a); the second in Fig. 6.8 (b). The situation in Fig. 6.8(b) is the one most likely to cause an error in pointing when the focus plane is very close to the reticle plane. In this case, the objective is focussed for the individual's eye and the cross-hairs are very close to being clearly focussed, but appear to move slightly across the image of the objective if the observer moves his head slightly from side to side or up and down. It can be immediately recognized that slightly different pointings could be made depending on the position of the eye behind the eyepiece. The situation of Fig. 6.8 (a) is not so usual since the lack of focus of the objective is generally easy to perceive. It can be appreciated that it is important to carefully focus the eyepiece before sighting any objective.

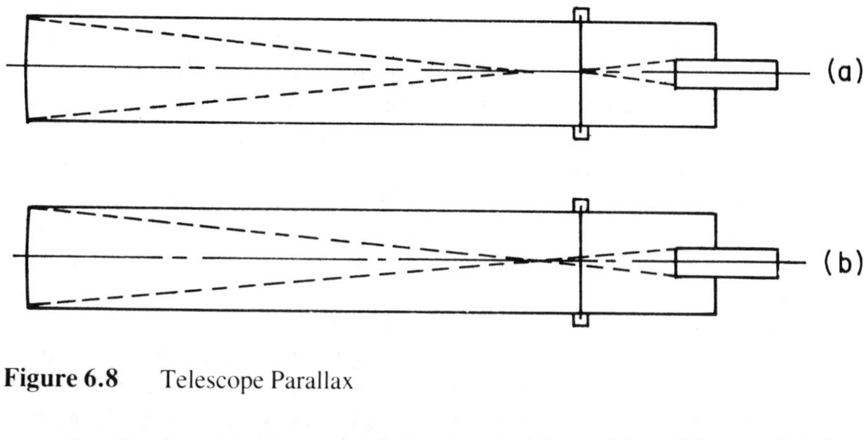

Figure 6.8 Telescope Parallax

Parallax is easy to detect simply by changing the position of the eye slightly. Experienced observers unconsciously check for this with nearly every pointing. The error is minimized simply by careful focussing of both lenses and by frequent checks for parallax when pointing. The error source is primarily personal although lighting and optical quality also influence the ability of the observer to focus properly. The error is unimportant if due care is taken to eliminate it.

Reading

Vernier instruments are generally assumed to have a random error in reading of ½ the least count. For example, a 30-second vernier system would be expected to have a reading error of ± 15 seconds. This error occurs with each reading or initial setting to zero; therefore, there are always two occurrences in any single or repeated angle series. Analysis of the propagation of this error will be included with the discussion of optical repeating theodolites.

Optical theodolites have various reading systems. There is some trend toward digital or direct read-out systems, but typical micrometer knob setting systems will undoubtedly remain as the most common design for the foreseeable future. The scale systems vary among models. Common designs for direction theodolites were shown in Figure 4.1 (a) and 4.1 (b). More examples are shown in Figure 6.9. These illustrations show what the observer sees when looking through the optical microscope. For the system in Figure 4.1 (a), the scale carrying the number 274 moves opposite the one carrying the number 94. When marks are all aligned using the micrometer knob, the reading is made. The one space between the 94 and 274 means 10 minutes. The remainder of the minutes and the seconds are read from the bottom scale where minutes and seconds appear. A study of the figure should reveal how the final value was observed. Figure 4.1(b), which is also 6.9(c), is a newer design. Marks are aligned in the top scale, using the micrometer knob, then the 94° 10′ read directly and added to the 2′ 44″ on the lower scale. Figure 6.9(d) is a scale for a repetition theodolite. The micrometer knob is used to align a number mark (327 in the figure) between two lines, at which time the minutes and seconds are read directly. Figure 6.9(b) is the most direct-reading of the group. It has no micrometer knob and is the closest scale system, in principle, to that of a vernier theodolite. It is not as important for this discussion to fully comprehend how to manipulate the knobs or take a reading. The exact techniques for each instrument are adequately covered in their respective manuals. The main point to be made here is that the reading error of an optical theodolite is comprised of two things: (1) the personal error in aligning markings on one scale or the other by rotating the micrometer knob, (2) the personal error in reading the divisions on the minutes or seconds scale after the marks are aligned. These two personal errors need not be separated when analyzing the reading error. Hereafter when "reading error" of an optical theodolite is mentioned, it will be assumed to be the composite error from both sources. It should be noted that setting the reading to zero on a repetition instrument includes an error from both sources.

The reading error source is primarily personal, but it is influenced by such natural and instrumental sources as the least count, the optical qualities of the system, and the illumination of the scales. The reading error is primarily random in nature. Any systematic personal error tendency of an observer to read "high" or "low" would tend to cancel in the process of measuring.

Theodolites with the coincidence system of reading, utilizing a micrometer knob, provide a means to isolate the reading error from other errors and to estimate it statistically. Section 4-4 explained how this is done. In that section the standard deviation of a single reading was computed for three different instruments. Let us next analyze the propagation of this error in angle measurement.

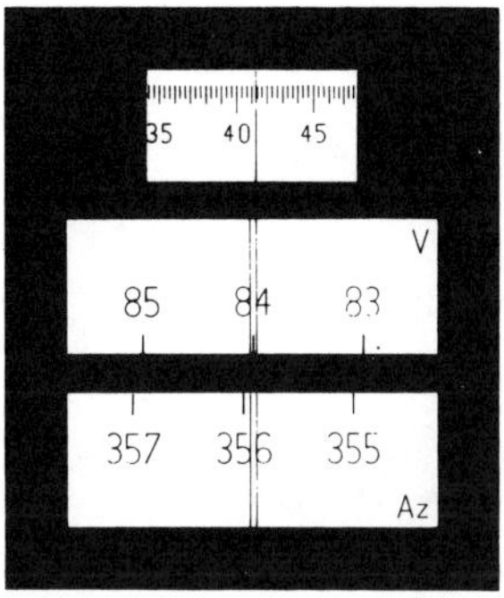

360° vertical circle reading 84° 41′ 15″

WILD TIA
(Prior to 1975)

(a)

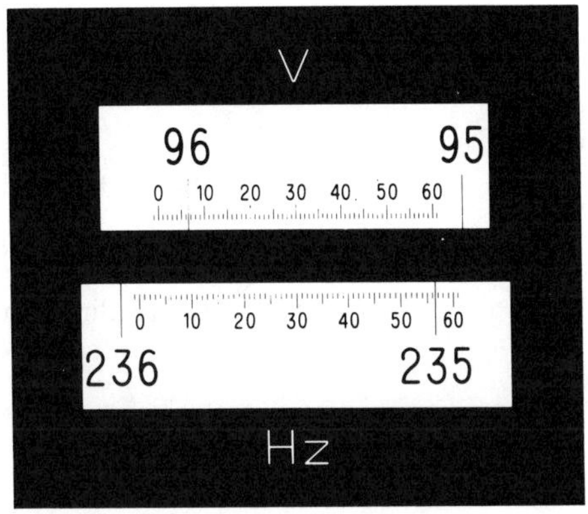

Circle reading: Vertical 96°06.5′
Horizontal 235°56.4′

WILD TI6

(b)

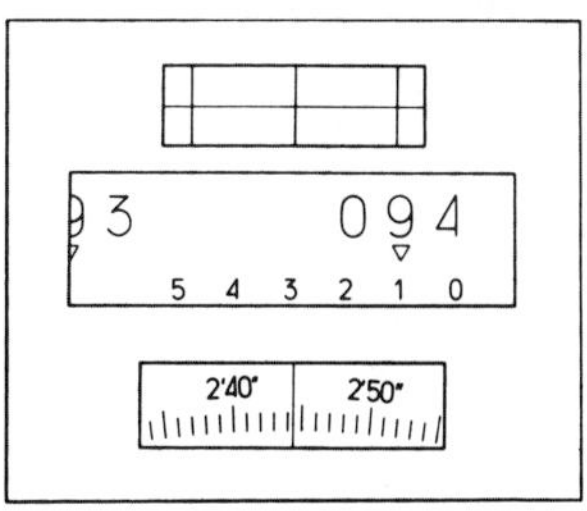

360° reading: Horizontal circle 94°12′44″

WILD T2
Direction System - See Fig. 4.1 for more

(c)

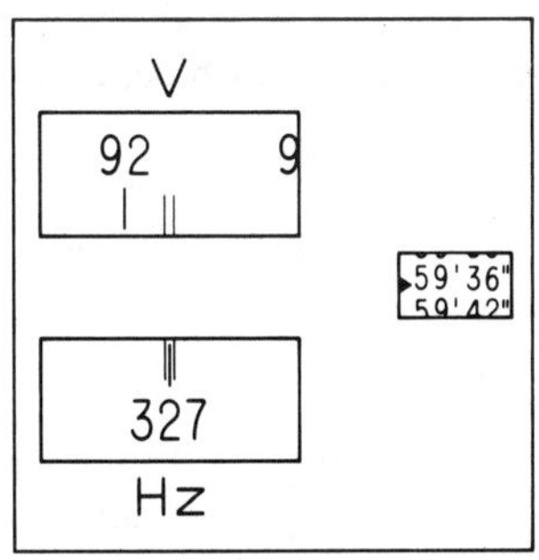

360° reading: Horizontal circle 327°59′ 36″

WILD T I

(d)

Figure 6.9 Repetition Theodolite Reading Scales (Courtesy of Wild Heerbrugg Instruments, Inc., Farmingdale, N.Y.)

A single horizontal angle is always, of necessity, comprised of two readings, one on the backsight and one on the foresight. The reading error occurs with each observed direction. Since the angle is the difference between readings, the errors propagate as random errors in addition.

$$\sigma_{\alpha_r} = \sqrt{\sigma_{r_1}^{\,2} + \sigma_{r_2}^{\,2}}$$

where

σ_{α_r} = the reading error in angle, α

σ_{r_1} = the reading error to target 1

σ_{r_2} = the reading error to target 2.

If,

$$\sigma_{r_1} = \sigma_{r_2} = \sigma_r$$

Then, $\sigma_{\alpha_r} = \sigma_r\sqrt{2}$ for a single angle measurement.

For either a vernier or optical type repetition instrument, there are only two occurrences of the reading error–one when it is zeroed, and one at the final reading of the accumulated angle. Any intermediate reading is used only for a check and is not used in the calculation of the angle. Therefore, σ_{α_r} remains constant for the accumulated angle. But, both the accumulated value and the reading error are divided by the number of repetitions. That is,

$$\alpha_m = \frac{\alpha_s \pm \sigma_r\sqrt{2}}{n},$$

where

α_m = mean angle,
α_s = accumulated angle
n = number of repetitions.

Thus, for an angle by repetition,

$$\sigma_{\alpha_r} = \frac{\sigma_r\sqrt{2}}{n} \tag{6.3}$$

where σ_{α_r} = standard deviation in angle α, due to reading error, σ_r, for n repetitions with a repetition theodolite.

For a direction theodolite, a reading must be taken at each pointing, thus the error occurs twice for each angle measured in the series. The error in the sum of the differences in readings to targets 1 and 2 is

$$\sigma_s = \sigma_r\sqrt{2}\sqrt{n}$$

which follows the propagation of errors in a series theory. And,

$$\alpha_m = \frac{\alpha_s \pm \sigma_r \sqrt{2} \sqrt{n}}{n},$$

where

α_m = mean angle
α_s = sum of differences in readings to targets 1 and 2
n = number of single angles measured.

Thus, for a mean angle with a direction theodolite,

$$\sigma_{\alpha_r} = \frac{\sigma_r \sqrt{2}}{\sqrt{n}} \tag{6.4}$$

where

σ_{α_r} = standard deviation in angle, α, due to reading error , σ_r, for n repetitions with a direction theodolite.

The above serves to illustrate the advantages of measuring angles more than once. Particularly for a repetition instrument, the effect of the reading error is reduced by doing so. The reading error can, of course, be reduced by selecting an instrument yielding a lower value for σ_r. Other means to minimize this error are to assure that the scales are adequately illuminated, that the optical system is kept clean, that appropriate care is taken in reading, and that personnel are properly trained and have the basic skills required of instrument operators.

Pointing

This error, also sometimes called the sighting error, results from pointing to one side or the other of the target center. This has nothing to do with target centering errors–only the error in perceiving the target's center, wherever it may be positioned. The pointing error is a function of the observer's care and ability, the width of the cross-hair, the target size and distinctness, the telescope power and optical qualities (particularly *definition*–see Section 3-1), and atmospheric conditions such as available light, heat waves, wind, haze. It is considered to be primarily a personal, random error, although influenced by the instrumental and natural factors mentioned. As can be determined through tests to be described, given all conditions the same except different instruments, higher pointing errors occur with some instruments primarily because of variation in definition.

An elementary appreciation of the significance of pointing error can be gained through Figure 6.10. The angular pointing error to a target located at B, is

$$\epsilon_p = \pm \frac{d_p}{D} \quad \text{(radians)} \tag{6.5}$$

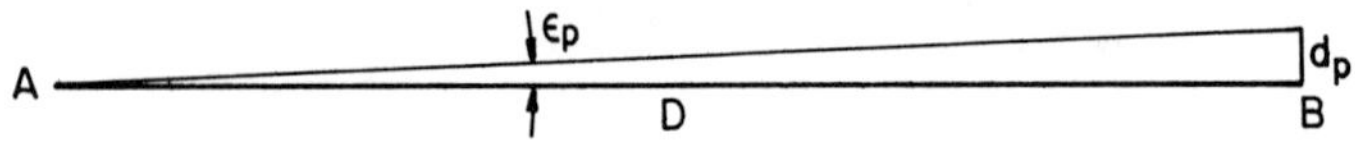

Figure 6.10 Pointing Error Effect on Angles

where

d_p = the linear pointing error to the target
D = distance from theodolite to target.

Reasonable estimates of likely pointing errors can be made from this relationship by using judgment to estimate d_p and D. For a vernier theodolite, this method must be used to estimate the error.

Theodolites with the coincidence reading system provide a means to determine the pointing error statistically. The procedure is similar to that used to determine the reading error except that an actual pointing is made to a target. A well defined target is selected and a pointing made. The reading is recorded. Next, the pointing is moved slightly, then realigned on the same target and another reading recorded. This is repeated enough times to have a sample large enough to compute a reliable value for the standard deviation in reading and pointing. It should be obvious that the value computed is a combined reading and pointing error. To isolate the pointing error, if desired, it is necessary to use data from reading error tests of the same instrument. Since it is assumed that the combined error is larger than the reading error by itself and that the two are added according to random error theory,

$$\sigma_p = \sqrt{\sigma_{rp}^2 - \sigma_r^2} \quad ,$$

where

σ_p = standard deviation in a single pointing
σ_{rp} = standard deviation in reading and pointing
σ_r = standard deviation in reading as determined by test.

Whether the theodolite is a repetition or a direction type, a pointing must be made to each target for each repetition. Therefore, the error occurs two times for each of the angles. The manner of occurrence of the pointing errors is identical to the reading error of a direction theodolite. Borrowing from the previous derivations for that error and skipping a few steps, it can be shown that, for either a repetition or direction instrument,

$$\sigma_{\alpha_p} = \frac{\sigma_p \sqrt{2}}{\sqrt{n}} \tag{6.6}$$

where

σ_{α_p} = standard deviation in angle, α, due to pointing error σ_p, for n repetitions.

If an analysis of both the reading and pointing error effect on angular uncertainty was being made for a repetition instrument, it is necessary to separate the two errors since one propagates inversely with n and the other with $\sqrt{n}$. However, except for academic or research purposes, there is no need to separate the reading and pointing error of a direction theodolite since they both propagate inversely with $\sqrt{n}$. For practical analyses, the value σ_{rp} can be used as determined from that test, whereby,

$$\sigma_{\alpha_{rp}} = \frac{\sigma_{rp}\sqrt{2}}{\sqrt{n}} \tag{6.7}$$

where

$\sigma_{\alpha_{rp}}$ = standard deviation in angle, α , due to combined reading and pointing error, σ_{rp} , for n repetitions.

Examples of analysis of reading and pointing error will be solved at the end of this section.

The error in an angle due to pointing error can be reduced by increasing the number of repetitions, as illustrated in the formulas. This error can also be reduced by selecting an instrument yielding a lower σ_p. Care in observing, such as careful focussing and observation of the target's center is important. Training and skill of the observer are factors in minimizing this error. Selecting targets with clear and distinct patterns is important. Vertical targets such as range poles, taping pins, or pencils should be selected so that they appear only slightly larger than the cross-hair width. They shouldn't be so narrow as to be completely covered by the cross-hair, nor should they appear more than about twice the width of the cross-hair. Using range poles at close range, although perhaps carefully centered on the ground point and having a sharp pointed end, causes large pointing uncertainties because of difficulty in estimating the pole's center. "Well defined" targets, are those that are of a size and pattern appropriate to the distance, as discussed at the end of Section 3-3. Pointing errors are minimized by avoiding adverse natural conditions such as darkness (unless lighted targets are used), heat waves, haze, and high wind. Heat waves cause the objective to be blurred and appear to dance in the field of view. Wind causes vibration of the instrument and possibly also the target. Vibration of the instrument is magnified through the telescope. Longer sight distances have little effect on pointing errors if targets are "well defined" and if the distance isn't so great as to diminish the clarity due to atmospheric conditions. This has been proved through tests of various theodolites using well defined targets at different distances.

Target Centering

This error results from sighting to a point on a target that is positioned in a vertical plane other than the one through the ground point. This can be caused by a number of factors, the personal error in centering or plumbing the target over the station being the primary source. Other factors influencing this error are instrumental errors such as a bent range pole, the tip of the range pole being off center with the axis of the pole, and maladjustment in the optical plummet of a tripod mounted geodetic target. Although random in nature, the error is systematic in effect unless targets are checked and centered with each pointing. In other words, once the targets are set over the stations, the same angular error results in the mean of several repetitions as in a single angle.

The angular error in a direction due to target misalignment is a function of the linear magnitude of the error, the sight distance, and the direction of the error with respect to the line, the maximum error occurring when the target is misaligned in a direction perpendicular to the line. The maximum angular error in direction 1 and direction 2 would be

$$\epsilon_{t_1} = \pm \frac{d_1}{D_1} \qquad \epsilon_{t_2} = \pm \frac{d_2}{D_2}$$

where

ϵ_{t_1} , ϵ_{t_2} = angular error to directions 1 and 2, respectively, in radians.

d_1 , d_2 = target centering errors, perpendicular to each line, 1 and 2.

D_1 , D_2 = sight distances 1 and 2.

The error in an angle, α, would be

$$\epsilon_{\alpha_t} = \pm \sqrt{\epsilon_{t_1}^2 + \epsilon_{t_2}^2} \tag{6.8}$$

Values for d_1 and d_2 are estimates, made by judgment.

Effects of target centering errors are minimized by careful centering and maintaining targets in good adjustment. The most effective way to reduce the effect is to keep sight distances as long as possible. Given equal care in centering, it is seen that the angular errors are reduced in inverse proportion to the distance. This is one elementary principle of planning field surveys–avoid short sight distances whenever possible.

Theodolite Centering

This error is similar in nature to the target centering error. It is primarily personal, being the observer's inability to use the optical plummet or plumb bob with exactness at the theodolite station. It can also be caused partly by the maladjustment of the optical plummet or other device used for centering over a station. Like target centering, the effect on an angle is systematic although the nature of the error is random. Once centered, the theodolite measures the same erroneous angle with each observation in a series. Therefore, the mean angle contains the same error as a single angle and the number of repetitions has no effect on reducing it.

The angular error in a direction due to the instrument centering error is a function of the same variables as the target centering error. But, the maximum angular error between two directions occurs when the error in centering lies along the bisector of the angle being measured, not when it is perpendicular to one of the lines. The maximum angular error is depicted in Fig. 6.11. In this figure,

$$\epsilon_{\alpha_c} = \frac{\pm\ d_c\ D_3}{D_1 D_2 \sqrt{2}} \tag{6.9a}$$

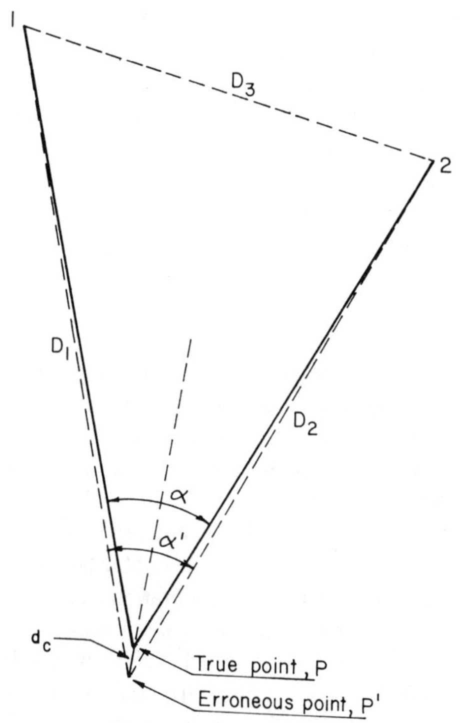

Figure 6.11 Theodolite Centering Error Effect on Angles

where

$$\begin{aligned}
\epsilon_{\alpha_c} &= \text{maximum angular error, } \alpha' - \alpha \text{ , in radians} \\
\alpha &= \text{true angle} \\
\alpha' &= \text{angle observed at erroneous station, P}' \\
d_c &= \text{linear centering error (eccentricity)} \\
D_1, D_2 &= \text{distance to targets 1 and 2, respectively} \\
D_3 &= \text{distance between targets 1 and 2.}
\end{aligned}$$

The value for d_c is an estimate, made by judgment. The derivation of this formula is left as a student exercise.

A simplified form of the equation can be derived when sight distances are nearly equal. If $D_1 = D_2 = D$,

$$D_3 = 2D \sin \alpha/2 \text{ .}$$

Substituting this into equation 6.9a yields

$$\epsilon_{\alpha_c} = \pm \frac{d_c \sqrt{2} \sin \alpha/2}{D} \tag{6.9 b}$$

It is seen from either formula that the angular error is directly proportional to D_3 which is larger for larger angles, or $\sin \alpha/2$ which also increases as α increases. This means that the error is small for small angles and approaches a maximum as the angle approaches 180°. The advantage of long sight distances becomes apparent when it is seen that increases in either D_1 or D_2 will decrease the angular error due to any given d_c.

Effects of instrument centering are minimized through care in centering, maintaining plummet devices in adjustment and keeping sight distances as long as possible.

Bubble Centering

The effect of personal errors in bubble centering is similar to that of maladjustment of the horizontal axis since any deviation in the inclination of the bubble axis changes the inclination of the horizontal axis the same amount. However, error effects are mostly random since the operator attempts to keep the bubble centered and will have an error which is as likely to incline the axis one direction as the other. The ability to keep the bubble centered is influenced by the clarity of markings on the bubble, the sensitivity of the bubble (see Section 3-2), and natural factors such as ground conditions, wind, vibrations, and temperature changes. A more sensitive bubble is more difficult to keep centered. But, the greater the sensitivity, the higher the precision, so difficulty in keeping centered doesn't adversely affect results. Settlement of the tripod is caused by poor set-up, soft or thawing ground, and rough handling of the instrument during measurement.

An equation has been derived for effects of inclination of the horziontal axis:

$$\mathbf{E}_H = \mathbf{E}_h \tan \gamma \tag{6.2}$$

Recall from Section 3-2 that bubble sensitivity is the angle subtended at the center of the circle of which the upper tube surface is a segment by a change in position of one division of the bubble (see Fig. 3.3). Note that if the bubble was off center one division when sighting in one direction, the angle E_h in Equ. 6.2 equals the sensitivity, μ. The uncertainty in the pointing to a target is

$$\epsilon_b = \pm f_d \mu \tan \gamma \tag{6.10}$$

where

ϵ_b = angular error in the pointing due to bubble being off center, in seconds,
f_d = the fraction of a division the bubble is estimated to be off, plus or minus,
μ = bubble sensitivity, in seconds,
γ = vertical angle.

Instrument Adjustments

At first glance, this may seem out of place under a heading of personal errors until it is realized that every instrument adjustment is made with the aid of a screwdriver or adjusting pin operated by a person! Every reading or observation of error in any of the instrument axes is judged by a human. When the adjustment is made, a person makes a judgment as to when to stop turning the screwdriver and declare the instrument to be adjusted. Furthermore, the instruments are not manufactured perfectly which can be traced to the inability of mortals to design and build the molds or turn the lathes with absolute perfection. Thus, no adjustment is ever perfect. There will always be a certain plus or minus error in all of the axes alignments. Recall that the tape length error is a systematic correction, but it has an uncertainty due to calibration of the tape. Theodolite axes will have uncertainties in alignment for similar reasons. Evaluation of the effects of these uncertainties on horizontal angles can be made by assuming a value (or using the instrument manufacturer's or repair technician's estimate) for the uncertainty in the variables in the error formulas for instrumental errors, then solving the equations to propagate the random errors.

Normally, uncertainties in horizontal angles due to this category of personal error would be negligible.

6-4 NATURAL ERRORS IN HORIZONTAL ANGLES

Wind causes the instrument to vibrate which affects precision in pointing. Strong gusts can actually shift the line of sight slightly and move targets. Wind also causes the bubble to change its position slightly, thus being off-center.

Temperature changes, especially as a result of radiant heat from the sun cause fluids in bubble tubes to expand or contract which shifts the ends of the bubble. Cold temperatures sometimes cause the leveling screws and other motions of the instrument to be stiff.

Heat waves, present in hot weather when the sky is clear, cause blurred images, especially for long sight distances.

Haze, whether natural or from industry or exhausts, decreases the effective sight distance and obscures the image at any distance.

Ground conditions such as thawing or marshy ground cause difficulty in achieving a firm tripod set-up. A slowly settling instrument causes the bubbles to constantly change centering, which in turn causes slight shifts in pointing and instrument centering.

Earth curvature and *atmospheric refraction* are factors affecting vertical angles only. These effects will be discussed in the next chapter on elevation measurements.

Horizontal refraction affects horizontal angles. This is the bending of the line of sight horizontally, which can occur when it passes very close to large objects such as buildings, especially if they are radiating heat. This can cause an erroneous pointing.

Effects of natural errors are difficult to quantify or derive formulas for evaluation. Some were discussed with personal errors since many influence the effect of personal errors. In general, the best way to deal with natural errors is to be ever aware of their occurrence so as to minimize their effects. Avoiding measuring under certain adverse conditions is the most obvious way to prevent their occurrence. Taking measures to minimize them during observation is the next best way. For example, simply realizing that the bubbles will drift when the ground is frozen should cause the operator to pay extra attention to their centering. Or, realizing that the line of sight may shift during a gust of wind should cause the observer to take extra care in pointing under these conditions. A large umbrella to shade the instrument will reduce effects of radiant heat on the bubble fluids.

6-5 DEALING WITH SYSTEMATIC ERRORS IN HORIZONTAL ANGLES

As has been mentioned, most instrumental errors are systematic. Personal and natural errors are primarily random. Systematic errors are normally compensated through the *principle of reversion* (measuring an equal number of angles with the telescope erect as with it inverted). Other systematic errors, caused by factors other than instrument maladjustment, can be corrected by computation. For example, if it was necessary to erect the target or the instrument over a point other than the true one, the true angle could be calculated. The eccentricity, its direction with respect to the line's direction, the sight distance, and the horizontal angle measured with the instrument or target over the nearby point would be known. The computation involves simple geometric relationships and can easily be made. The magnitude of systematic errors caused by instrument maladjustment can be evaluated using Equations 6.1 and 6.2 if the angle of maladjustment is known. This is a useful exercise to determine the importance of inverting the telescope for an equal number of observations and also illustrates that for certain types of work, such errors may even be negligible so that the double angle procedure may be unnecessary in such instances.

6-6 PROPAGATION OF RANDOM ERRORS IN HORIZONTAL ANGLES

Determining Effects from Individual Sources

Horizontal angle uncertainty can be estimated by assuming reasonable values for all personal errors, then applying the several error formulas derived. The total effect on the angle is the sum of the separate effects, added according to the law of addition of random errors. There are many variables to be considered for any given angle. For this reason, it is not wise to simply assume an uncertainty per station, although this expedient method is generally used in most specifications. The following examples will illustrate how variation in adjustment, bubble centering, bubble sensitivity, eccentricity of instrument and target, reading and pointing errors, sight length, vertical angle, number of repetitions, and type of instrument, all influence the uncertainty in any horizontal angle. Dealing with each source individually yields better estimates of uncertainty than assuming a combined average from all sources and also provides a means to balance or adjust angles of a closed traverse by applying weights based on the individual uncertainties.

Detailed calculations of the type to be illustrated are somewhat tedious and perhaps impractical in private practice if made with a calculator. However, electronic computers can easily be programmed to make such estimates quickly and inexpensively. The adjustments of the angles can be made in the same set of program statements, all of which should result in more precise final adjusted directions and point positions than if angles are assumed to have the same uncertainty.

Example 6-1. Given: The first instrument tested in Section 4-4 yielding a standard deviation in reading, $\sigma_r = \pm 0.65''$. This same instrument, when tested for combined reading and pointing errors, yielded a value $\sigma_{rp} = \pm 1.2$ seconds. What is the uncertainty in a single angle measured with this theodolite, from only the reading and pointing errors?

Solution: Since both the reading and pointing error occurs with each pointing using this instrument, there is no need to consider the reading error separately. The error in an angle α occurs twice–once for each pointing

$$\sigma_{\alpha_{rp}} = \pm \sigma_{rp} \sqrt{2}$$

from which

$$\sigma_{\alpha_{rp}} = \pm 1.7'' .$$

Example 6-2. Given: The instrument of Example 6-1 and 8 repetitions of the angle. What is the uncertainty in the mean angle from only the reading and pointing error?

Solution:

$$\sigma_{\alpha_{rp}} = \frac{\sigma_{rp}\sqrt{2}}{\sqrt{n}} \tag{6.7}$$

where

$$n = 8 \text{ repetitions}$$

from which

$$\sigma_{\alpha_{rp}} = \pm 0.6'' .$$

Example 6-3. Given: A repetition theodolite was tested using the procedures previously discussed for reading and pointing error. The results were $\sigma_{rp} = \pm 3.9''$ and $\sigma_r = \pm 3.0''$. What is the uncertainty in the mean of an accumulation of 6 repetitions of a horizontal angle?

Solution: First, the pointing error must be separated from the combined error since it will propagate different than the reading error.

$$\sigma_p = \sqrt{\sigma_{rp}^2 - \sigma_r^2} = \sqrt{3.9^2 - 3.0^2} = \pm 2.5''$$

From Section 6-3,

$$\sigma_{\alpha_r} = \frac{\sigma_r\sqrt{2}}{n} \tag{6.3}$$

and

$$\sigma_{\alpha_p} = \frac{\sigma_p\sqrt{2}}{\sqrt{n}} \tag{6.6}$$

where

$$n = 6$$

from which

$$\sigma_{\alpha_r} = 0.71$$
$$\sigma_{\alpha_p} = 1.44$$

the combined error follows the law of addition of random errors

$$\sigma_{\alpha_{rp}} = \sqrt{0.71^2 + 1.44^2} = \pm 1.6''$$

Example 6-4. Given: Assume a target centering error of ± 0.1 inches for each of two targets, one at 1,000 feet and the other at 2,000 feet from the theodolite. What is the maximum uncertainty in a single horizontal angle between these two targets?

Solution: The maximum angular error would result when the target centering error is perpendicular to the direction of each pointing. From Section 6-3,

$$\epsilon_{t_1} = \frac{\pm d_1}{D_1} = \frac{\pm 0.1}{1{,}000 \times 12}\, 206{,}265 = \pm 1.7''$$

$$\epsilon_{t_2} = \frac{\pm d_2}{D_2} = \frac{\pm 0.1}{2{,}000 \times 12}\, 206{,}265 = \pm 0.9''$$

$$\epsilon_{\alpha_t} = \sqrt{\epsilon_{t_1}^{\,2} + \epsilon_{t_2}^{\,2}} = \pm 1.9'' \qquad (6.8)$$

Example 6-5. Given: The same target centering error in Example 6-4 with sight distances to targets 1 and 2 of 50 feet and 80 feet and four repetitions of the angle. Compute the maximum uncertainty in the mean horizontal angle between these two targets.

Solution:

$$\epsilon_{t_1} = \frac{\pm d_1}{D_1} = \frac{\pm 0.1}{50 \times 12}\, 206{,}265 = \pm 34.4'',$$

$$\epsilon_{t_2} = \frac{\pm d_2}{D_2} = \frac{\pm 0.1}{80 \times 12}\, 206{,}265 = \pm 21.5'',$$

$$\epsilon_{\alpha_t} = \sqrt{34.4^2 + 21.5^2} = \pm 40.6'' \text{ (say } \pm 40'')$$

It is noted that repetitions of the angle have no effect on the answer. The same error occurs regardless of the repetitions since the same erroneous angle is measured each time. A comparison of the answers to Examples 6-4 and 6-5 demonstrates the effect of sight distance on horizontal angle uncertainty. Extreme care must be taken in target centering when sight distances are short. Furthermore, control traverse lines should be as long as possible for the reasons illustrated by these examples.

Example 6-6. Given: The data of Example 6-4 and an estimated instrument centering error of ± 0.1 inches with an angle of approximately 30°. What is the maximum uncertainty in the angle from both target and instrument centering?

Solution: For this problem, D_3 can be determined to sufficient precision from a plot of D_1, D_2, and α by scaling (See Fig. 6.11). Alternatively, D_3 can be computed using the law of cosines.

$$\epsilon_{\alpha_c} = \frac{\pm d_c D_3}{D_1 D_2 \sqrt{2}} \qquad (6.9a)$$

where

$$D_1 = 1{,}000 \text{ feet}$$
$$D_2 = 2{,}000 \text{ feet}$$
$$D_3 = 1{,}240 \text{ feet}$$
$$d_c = \pm\ 0.1 \text{ inch}$$

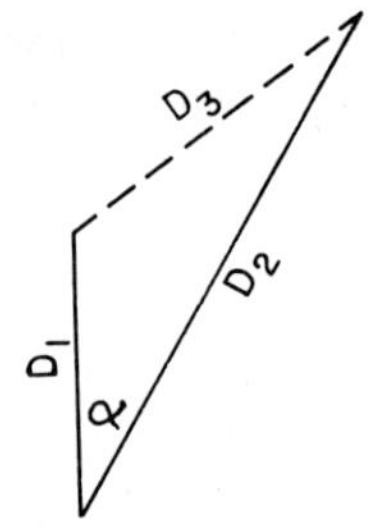

After converting d_c to feet and multiplying by the conversion factor for seconds per radian,

$$\epsilon_{\alpha_c} = \pm\ 0.75''$$

Thus the error from all three sources,

$$\epsilon_\alpha = \sqrt{\epsilon_{t_1}^{\ 2} + \epsilon_{t_2}^{\ 2} + \epsilon_{\alpha_c}^{\ 2}} = \sqrt{1.9^2 + 0.75^2} = \pm 2.0''$$

Example 6-7. Given: The vertical angle to target 1 is +2° 14′. The vertical angle to target 2 is −5°30′. The plate bubble of the theodolite has a sensitivity of 20 seconds. An estimate of the bubble centering uncertainty is ± ½ of the bubble division width. What is the uncertainty in a double angle due to bubble centering error?

Solution: From Section 6-3,

$$\epsilon_{b_1} = \pm\ f_d \mu \tan \gamma_1 \tag{6.10}$$

and

$$\epsilon_{b_2} = \pm\ f_d \mu \tan \gamma_2$$

where

$$f_d = \pm ½$$
$$\mu = 20 \text{ seconds}$$
$$\gamma_1 = +\ 2°\ 14'$$
$$\gamma_2 = -\ 5°\ 30'$$

from which

$$\epsilon_{b_1} = \pm\ 0.39''$$
$$\epsilon_{b_2} = \pm\ 0.96''$$

The total uncertainty from this source, for the double angle is

$$\epsilon_{\alpha_b} = \frac{\sqrt{\epsilon_{b_1}^2 + \epsilon_{b_2}^2 + \epsilon_{b_1}^2 + \epsilon_{b_2}^2}}{2}$$

$$= \frac{\sqrt{\epsilon_{b_1}^2 + \epsilon_{b_2}^2}}{\sqrt{2}} = \pm\ 0.73'' \text{ or } \pm\ 0.7''$$

It is noted that the error occurs with each pointing and is therefore purely random in effect. Thus, it is reduced according to the usual error of the mean formula.

Example 6-8. Given: The instrument of Example 6-1, the target centering errors and target distances of Example 6-4, the instrument centering error and horizontal angle of Example 6-6, and the plate bubble sensitivity, vertical angles, and bubble centering uncertainty of Example 6-7. Calculate the uncertainty in 3 sets of the horizontal angle, considering reading, pointing, target centering, instrument centering, and bubble centering errors.

Solution: First of all, it is noted that $n = 6$ since a ''set''consists of 2 angle measurements, one with the telescope direct and one with it inverted. This solution must consider that the reading and pointing errors were determined by a statistical test and the other errors by pure estimate. Therefore, assumptions must be made as to the *level of certainty* of the estimates. That is, were they only 68% certain? Or, perhaps only 50% certain? It would seem that a level of certainty of about 90% is reasonable to assume for the estimates. But, other levels could also be assumed since it is only an estimate. If 90% is assumed, then the σ value for the reading and pointing error must be converted to E_{90}. The calculations for all errors are as follows:

$$E_{\alpha_{rp}} = \frac{\pm 1.645\ \sigma_{rp}\ \sqrt{2}}{\sqrt{n}} = \frac{\pm\ 1.645\ (1.2)\ \sqrt{2}}{\sqrt{6}} = \pm\ 1.14''$$

(Note that this is Equ. 6.7 with E_{90} conversion)

$$E_{\alpha_t} = \pm\ 1.92'' \qquad \text{(from Ex. 6-4)}$$

$$E_{\alpha_c} = \pm\ 0.75'' \qquad \text{(from Ex. 6-6)}$$

$$E_{\alpha_B} = \frac{\sqrt{\epsilon_{B_1}^2 + \epsilon_{B_2}^2}}{\sqrt{6}} = \pm\ 0.42'' \qquad \text{(Ex. 6-7 data)}$$

$$E_{90_\alpha} = \sqrt{1.14^2 + 1.92^2 + 0.75^2 + 0.42^2} = \pm 2.4''$$

Analyzing Effects of Individual Errors

In Example 6-8, it is seen that all of the sources contributed to the total error to some extent. Had the instrument been less precise in reading, this error would have contributed most. Indeed, had it been a 30″ or 1′ transit, all of the other errors could have been neglected. If the angle had been larger, the instrument centering error would have been more significant. An angle of 150° instead of 30°, for example, would have caused this error to be ± 1.75″ instead of ± 0.75″. Longer sight distances would have decreased the effects from both instrument and target centering, but shorter distances would have increased their effects on the angle. Related to this point is the revelation that the precision of a 1-second instrument is wasted if sight distances are much shorter than those in this example unless target and instrument centering can be done much more precisely. If the terrain had been flat, the effect from the bubble centering would have been of no consequence, but there would have been more concern for this error in mountainous terrain or over short distances with large elevation differences.

It should be obvious that the total uncertainty in a horizontal angle varies with many changing conditions–among them being variations in the instrument precision, the number of repetitions, the terrain, the line lengths, the care in centering instruments and bubbles, and other factors. Therefore, the uncertainty at each station will vary. Even if the same repetitions, care, and instrument is used, other variables will change.

Expected Angular Closure

In a closed traverse, the sum of the interior angles is equal to $(n-2)$ 180°, plus or minus the error of closure. If the actual error of closure exceeded the expected error, one or more mistakes or poor work would be suspected. The surveyor would then need to find the mistakes or locations of poor work in the field before continuing to make use of the data.

$$E_{ca} = \Sigma \measuredangle\text{'s} - (n-2)\,180° \tag{6.11}$$

where

E_{ca} = actual error of closure
$\Sigma \measuredangle$'s = actual sum of the mean angles at each station
n = number of traverse stations.

$$E_{ce} = \sqrt{\epsilon_{\alpha_1}^{\,2} + \epsilon_{\alpha_2}^{\,2} + \cdots + \epsilon_{\alpha_n}^{\,2}} \tag{6.12}$$

where

E_{ce} = expected error of closure in a traverse of n sides
$\epsilon_{\alpha_1}, \epsilon_{\alpha_2}, \cdots, \epsilon_{\alpha_n}$ = total uncertainty at each station, $1, 2, \cdots, n$ according to estimates at each station.

The values for each ϵ_α might be computed much like the procedure in Example 6-8, an uncertainty thus being computed for each station. Alternatively, the average traverse situation might be assumed for line lengths, vertical angles, and other variables and an average uncertainty computed for each angle. This would not consider extreme cases at individual stations, but might suffice for an estimate. In this instance,

$$\epsilon_{\alpha_1} = \epsilon_{\alpha_2} = \cdots \epsilon_{\alpha_n} = \epsilon_\alpha$$

and from Equ. 6.12,

$$E_{ce} = \epsilon_\alpha \sqrt{n} \tag{6.13}$$

Example 6-9. Given: The horizontal interior angles of a traverse have been measured using a repetition theodolite having a reading error, $E_{90} = \pm 3.4$ seconds and a pointing error, $E_{90} = \pm 2.2$ seconds. The double angle procedure was used. Distances ranged from 150 to 300 meters, an average value of 200 meters being assumed. Vertical angles varied between 1 and 2 degrees. The 20-second plate bubble was kept centered ± ½ division. The traverse had 10 sides. Angles varied from 90° to 160°. It is estimated that the theodolite and targets were centered ± 2 mm. Targets of an appropriate size were used, resulting in well defined targets for optimum pointing. All error estimates are considered 90% certain. Neglecting any unaccounted natural or other errors, what would be the expected angular closure of this traverse, assuming average conditions at each station?

Solution: Let us consider each error source individually:

Reading —

$$\epsilon_{\alpha_r} = \frac{\epsilon_r \sqrt{2}}{n} = \frac{\pm 3.4\sqrt{2}}{2} = \pm 2.4'' \text{(from Equation 6.3)}$$

Pointing —

$$\epsilon_{\alpha_p} = \frac{\epsilon_p \sqrt{2}}{\sqrt{n}} = \frac{\pm 2.2\sqrt{2}}{\sqrt{2}} = \pm 2.2'' \quad \text{(from equation 6.6)}$$

Target Centering —

$$\epsilon_{t_1} = \epsilon_{t_2} = \frac{d}{D}\rho = \frac{2\text{mm} \times 206{,}265}{(200 \times 1000)\text{mm}} = \pm 2.06''$$

$$\epsilon_{\alpha_t} = \sqrt{\epsilon_{t_1}^2 + \epsilon_{t_2}^2} = 2.06\sqrt{2} = \pm 2.9'' \tag{6.8}$$

Instrument Centering — An average interior angle of 144° must be assumed.

$$\epsilon_{\alpha_c} = \frac{\pm d_c \sqrt{2} \sin \alpha/2}{D} \tag{6.9b}$$

$$= \frac{\pm\ 2\text{mm}\ \sqrt{2}\ \sin 72^\circ}{(200 \times 1000)\ \text{mm}}\ 206{,}265 = \pm 2.8''$$

Bubble Centering — Using an average value of 1.5° for the vertical angles yields, for one direction,

$$\epsilon_{b_1} = \pm\ f_d\mu \tan\gamma = \pm\ \tfrac{1}{2} \times 20''\ \tan 1.5^\circ = \pm\ 0.26'' \qquad (6.10)$$

and for both directions and the double angle,

$$\epsilon_{b_\alpha} = \epsilon_{b_1} \frac{\sqrt{2}}{\sqrt{2}} = \pm\ 0.26''$$

It is seen already that this error is negligible and will not be considered further in this example.

Total Station Error — For any one of the 10 stations, the expected error would be

$$\epsilon_\alpha = \sqrt{2.4^2 + 2.2^2 + 2.9^2 + 2.8^2} = \pm\ 5.2''$$

And, the expected error of closure for the traverse would be

$$E_{ce} = \pm\ \epsilon_\alpha \sqrt{n} = \pm\ 5.2\sqrt{10} = \pm\ 16.4'' \qquad (6.13)$$

Therefore, if the actual error of closure from Equ. 6.11 exceeded 16 seconds, the traverse angles would be considered faulty and the data not used further until the problem was corrected. As will be discussed in Chapters 9 and 10, it can be said that the traverse closure did not meet "specifications" and "adjustments" of the data would not be made if the expected closure was not achieved.

It should be mentioned that the above expected closure contains a 90% certainty if the estimates are accurate. Thus, it is fairly safe to assume that if the actual error exceeded this, there was a problem in the field. However, it should also be mentioned that some possible error sources were not evaluated. For example, minor natural errors such as settlement or wind could cause a misclosure rather than faulty work or mistake. For this reason, it may be fair to make the specification equal perhaps 20 or more seconds instead of 16 seconds just to give the benefit of the doubt to field personnel for unaccounted and unavoidable minor random errors and for conditions at some stations varying from average. More will be discussed on this point in Chapter 9.

Estimation of expected closure and evaluation of uncertainty is an important part of surveying measurement theory and practice. Such ability elevates the surveyor from craftsman or technician toward professional status. Such evaluation is the first step toward deriving specifications and controlling uncertainty, to be discussed in Chapter 9.

Problems - Chapter 6

6.1 Draw separate sketches of each of the axes of a theodolite showing their desired relationships with other axes.

6.2 A horizontal angle was measured between two targets from an instrument station. One of the targets had to be offset from the true station due to obstacles along the sight line. It was offset a distance of 2.0 feet perpendicular to and in a direction causing a larger angle reading. If this sight distance was 2,500 feet, what correction should be applied to the mean of the angle set observed?

6.3 Which instrument maladjustments are compensated when measuring an equal number of angles with the telescope direct as inverted?

6.4 Why must the bubble be kept centered when making any pointing for horizontal angle measurement?

6.5 In performing a test for line of sight adjustment of a theodolite as described by Figure 6.2, the points B and D were 1.2 cm apart. Distance CI was 140 meters. What is the adjustment error, in seconds, between the line of sight and horizontal axis?

6.6 Using the results of Problem 6.5, what would be the angular error in pointing to a target at 10° 30′ vertical angle?

6.7 For a test of inclination of the horizontal axis of a theodolite, the side of a building was used as described by Fig. 6.4. The theodolite was set up 100 feet from the building and a point "A" sighted at a vertical angle of 32° 38′. When the test was completed, BC was measured to be 0.32 feet. What is the adjustment error between the horizontal and vertical axes, in seconds?

6.8 Assume that a horizontal angle was measured 6 times with the unadjusted theodolite tested in Problem 6.7 but the operator did not invert the telescope between sets. The vertical angle to target 1 is −5° 12′ and to target 2 is +8° 46′. What would be the error in the mean angle due to inclination of the horizontal axis, in seconds?

Use the data below for Problems 6.9 through 6.15. Data are from reading and pointing error tests of a direction theodolite, taken as described in Chapter 4 and in this chapter.

Reading Test Data:					Reading & Pointing Combined Data				
27.5	29.2	30.8	28.1	30.6	42.0	44.4	48.0	46.0	43.5
28.5	27.7	29.2	28.9	29.0	45.7	41.7	46.1	41.9	45.3
30.5	28.9	29.8	26.5	27.5	42.2	43.9	39.5	44.7	42.5
28.6	27.8	29.1	28.0	25.2	42.3	40.2	42.1	42.0	47.2
29.2	28.3	29.0	29.5	28.1	45.3	43.7	43.0	39.2	42.3

6.9 For the above, plot histograms of each set of data using a class interval of approximately the σ value for each. Check for symmetry and randomness by inspecting the histograms and comparing the mean, the median, and the mode of each set.

6.10 Compute the standard deviation of a single value for each set of data. Identify these as σ_r and σ_{rp}, respectively, and further check for symmetry by counting the values in the σ range and E_{90} range.

6.11 Compute the error in pointing, σ_p for this instrument-observer combination.

6.12 Compute the 90% error due to reading and pointing for the mean of six repetitions of an angle using this instrument.

6.13 Using the pointing error computed in Problem 6.11, would this observer have the cross-hair on the range pole 95% of the time at a distance of 1,000 meters if the range pole is 3.0 cm wide? Show computations and sketch.

6.14 Recompute the σ_r and σ_{rp} values by rounding each observation to the nearest second before computing the σ values. Comment on the difference between these values and those of Problem 6.10.

6.15 Would you say that angles measured with this instrument are significant to the nearest second for most situations? Why or why not?

6.16 Suppose a repetition theodolite has been tested with values of $\sigma_r = \pm 3.0''$ and $\sigma_{rp} = \pm 5.0''$. What is the 99% error in 10 repetitions of the angle considering only reading and pointing errors?

6.17 Show by sketch and application of Equ. 6.9 that the error due to instrument centering is maximum when the horizontal angle is 180°.

6.18 How close should plate bubbles be kept centered in terrain with vertical angles averaging 4° if errors from this source are not to exceed those due to reading and pointing? Assume a plate bubble sensitivity of 20″, the instrument of Problems 6.9 - 6.15 and 90% certainty levels.

6.19 Horizontal angles were measured for a 15-sided open traverse between two azimuth marks 10 Kilometers apart over terrain having maximum elevation differences between stations of 20 meters using a direction theodolite having a reading error, $\sigma_r = \pm 0.8''$ and pointing error, $\sigma_p = \pm 1.5''$, and plate bubble sensitivity = 20″. Assume a bubble centering error of ± ½ division. Using a 95% certainty level, what is the expected error of closure of this traverse, assuming the azimuths at the terminal points are without error?

6.20 An angle is to be established to ± 5 seconds using a repetition theodolite. A single angle of the desired value was laid off from the base line by estimating ¼ of the 20″ circle division. The distance to the point set is 1860 feet. From tests of the instrument and subsquent calculations, it is known that 12 repetitions are actually needed to have this angle to the desired 5-second precision. An accumulated value of 182° 46′ 35″ was observed for 12 repetitions to the point established by the single angle measurement. If the desired value of the angle is 15° 13′ 50″, what should be done to establish the final station?

Additional Reading References - Chapter 6

1. Blachut, T. J., Chrzanowski, A., and Saastamoinen, J.H.. *Urban Surveying and Mapping*. New York: Springer-Verlag, 1979, Chapter 3 (on angle measurement).

2. Brinker, Russell C. and Wolf, Paul R. *Elementary Surveying*. Sixth Ed. New York: Harper and Row, 1977, Chapter 8.

3. Moffit, Francis H. and Bouchard, Harry. *Surveying*. Seventh Ed. New York: Harper and Row, 1981, Chapter 6, 19.

CHAPTER 7

ANALYSIS OF ELEVATION MEASUREMENTS

7-1 ANALYSIS OF DIFFERENTIAL LEVELING MEASUREMENTS

Basic definitions related to vertical distances were given in Section 1-6. Instruments, methods and field notes for differential leveling were discussed in Section 3-5, where additional definitions were given. As can be seen from a review of those sections, actual execution of differential levels and the basic theory is not complex. Likewise, as will be seen, analysis of the measurements is not as complex as for angle and distance measurements.

Instrumental Errors

Line of sight. If the line of sight is inclined with respect to the horizontal, an error will occur in a rod reading, proportional to the amount of inclination and the sight distance. The error is systematic and is revealed by an instrument test. The error cancels in the leveling process if backsight and foresight distances are equal. If there is an accumulated error in the difference in the sums of these distances between points, a correction must be made if the line of sight is not in proper adjustment. It is good practice to adjust the line of sight as part of the instrument testing process and avoid any corrections or random errors due to accumulated errors or uncertainty in sight distances.

It is noted that the line of sight error can occur whether the instrument has an automatic compensator or a level bubble. For an automatic level, the error would be present if the line of sight was not horizontal when the compensator was swinging free. For a tilting, dumpy, or wye level, the error would occur if the line of sight was not parallel with the bubble axis. The compensator must be free or the bubble tube centered when rod readings are taken in the instrument test to be described.

The test for line of sight inclination is illustrated in Fig. 7.1. The level is set up midway between two firm and stable points, A and B. It is leveled and rod readings a_1 and b_1 observed with the rod held at points A and B, respectively. The elevation difference between points A and B is without error as determined by $a_1 - b_1$ because any line of sight error cancels, the sight distances being equal. Next, the level is set up very close to either point A or B, leveled, and new rod readings observed on both points. In Fig. 7.1, the new reading b_2 at B is essentially without error due to line of sight inclination since the instrument is so close to the rod. The error is revealed by taking the reading a_2 on a rod held at point A.

In Fig. 7.1, the solid line represents a true horizontal line through the telescope and the dashed line represents the inclined line of sight, which could be

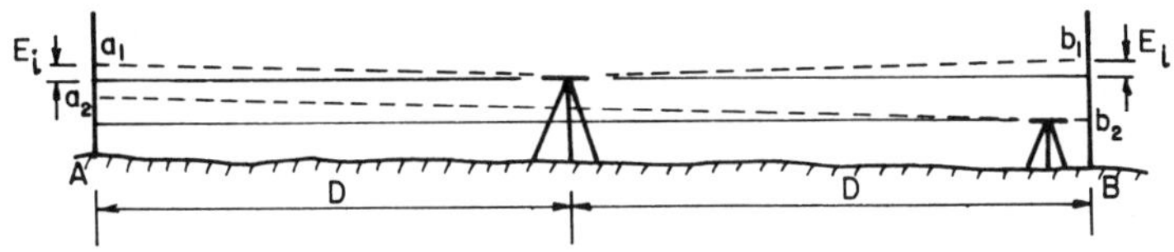

Figure 7.1 Horizontal Line of Sight Test

either up or down. The true elevation difference is

$$\Delta \text{ Elev} = a_1 - b_1 = (a_2 - E_i) - b_2$$

where

E_i = the inclination error per distance 2D, and a_1, b_1, a_2, b_2 are rod readings as have been explained.

$$E_i = (a_2 - b_2) - (a_1 - b_1).$$

The sign of E_i is positive if the line of sight is inclined upward and negative if downward. The line of sight can be adjusted by moving the horizontal hair to read

$$a = (a_1 - b_1) + b_2$$

with the instrument in its position near point B. This adjustment is performed according to specific instructions in instrument manuals.

It should be mentioned that, with some tilting levels, it is not necessary to test the line of sight with this "two-peg" test described above, as such instruments are provided with a means to rotate the telescope 180 degrees about the line of sight itself, thus enabling testing its inclination while pointing in one direction.

Bubble axis. As was discussed in Section 6-2 in reference to a theodolite, if the bubble axis is not perpendicular to the vertical axis, the bubble will move when the instrument is pointed in different directions. The same occurs with a level which has a bubble tube as a leveling mechanism. In leveling there is no error in a rod reading if the bubble is centered when the reading is observed, assuming the bubble axis is parallel with the line of sight. A lack of perpendicularity between bubble and vertical axis is merely a nuisance factor since the bubble may need to be releveled with each rod reading. An error would occur only if the operator neglected to center the bubble–and, this should be categorized as a mistake rather than error.

The test and adjustment for a dumpy or a wye level is the same as that described in Section 6-2. For a tilting level, the adjustment is similarly made.

Specific instructions in instrument manuals should be followed for each instrument.

On an automatic level, the only bubble is the circular bubble. Even an appreciable maladjustment in this bubble will not cause the compensator to bind. Its adjustment is easily made following instrument manuals for each model.

Cross-hairs. The adjustment of the cross-hairs of a theodolite was discussed in Section 6-2. Fig. 6.7 illustrates a reticle which is rotated. Of concern in leveling is the horizontal, rather than the vertical cross-hair. If the line of sight is in adjustment but the rod is read at a point on the cross-hairs other than the center, there would be an error in the reading if the cross-hair was not horizontal. The error would tend to cancel if the same location was read on the cross-hair each time, and would theoretically be zero if the vertical hair was centered on the rod at the point of reading. As a matter of good practice, the rod should be read near the center of the reticle image to render this error negligible. Furthermore, the reticle should be adjusted as described in Section 6-2 so as to avoid the error.

Rod length. An error in the rod length can occur in one of two ways: (1) an abrupt change at a specific place on the rod, (2) a systematic error distributed throughout the rod length.

The first type of rod error can occur if the bottom of the rod is worn, has something stuck to it, or has become damaged. It can also occur between rod joints when they are poorly fitting. It is unlikely to occur anywhere else on the rod. Any errors in rod readings caused by a constant and unchanging error at the bottom of the rod cancel in the leveling process. Errors caused by joint misfits do not cancel if readings above the joint are observed for some readings and below the joint for others. The error would accumulate when leveling up or down a slope since readings would be consistently above the joint in one direction and below it in the other. A rod can easily be checked by measuring across the joint with a steel tape.

The second type of rod length error is similar in nature and effect to a tape length error. It is analyzed in much the same way, the major difference being that rod length errors affect vertical rather than horizontal distances. When it is realized that a level rod is merely a rule which measures vertical distances in segments much like tape lengths, this is more easily comprehended. A ''segment'' is the backsight minus the foresight reading in each set-up. These differences (segments) are then added algebraically and the sum contains the sum of the rod length error effects much like the sum of tape length errors in horizontal distance measurements. Examples at the end of this section illustrate this and other points about the rod length error.

Automatic compensator. This error applies only to automatic levels. Its counterpart is the personal bubble centering error for instruments employing a bubble tube for leveling. Even if the line of sight of an automatic level is adjusted, it still has an uncertainty in pointing along the horizontal due to friction in the compensator movement and the overall precision of the mechanism in its ability to respond to gravity effects. The user has no way to check or estimate this error. The manufacturers generally list a ''compensator accuracy'' in the technical data for their instruments, given in seconds of arc for the uncertainty in direction of the line of sight with respect to the horizontal when the compensator is free. Thus, the effect on a rod reading

$$e_a = \pm c D \tag{7.1}$$

where

e_a = the error in a rod reading in sight distance, D,

c = the manufacturer's compensator accuracy, converted to radians.

This error is purely random in nature and effect on elevation differences in the differential leveling process.

Personal Errors

Plumbing rod. If the rod is not vertical, any reading will contain a positive error (be too high). A rod bubble, such as that illustrated in Fig. 3.34 is commonly used to plumb the rod. Another technique is to hold the rod loosely between the fingertips and note whether it tends to fall in one direction or the other. When the position is found where the rod does not fall in any direction, it is plumb and the rod operator then holds the rod firmly in that position. Still another technique is to rock or wave the rod slowly toward and away from the instrument. As the rod passes through the vertical, the instrument operator notes the reading, which will be the smallest observed. This latter technique has certain disadvantages in that it generally requires more observation time and there is a position where a rod with a flat bottom may actually yield a slightly lower reading than if plumb. The vertical distance effect of the rod being out of plumb is analogous to the horizontal distance effect of a tape being held on a slope, and is analyzed in a similar manner. Figure 7.2 illustrates the effect of the rod being out of plumb.

From an analysis similar to that for slope,

$$E_v = \frac{d_v^2}{2r} \tag{7.2}$$

where

E_v = the error in rod reading

d_v = the amount the rod is out of plumb at reading, r.

Although random in nature, the effect of this error on an individual rod reading is systematic since the error is always positive. However, the error is compensatory in the leveling process since it affects both backsight and foresight readings which are subtracted. It is seen that the error is a function of both the amount the rod is out of plumb and the value of the rod reading. For a given

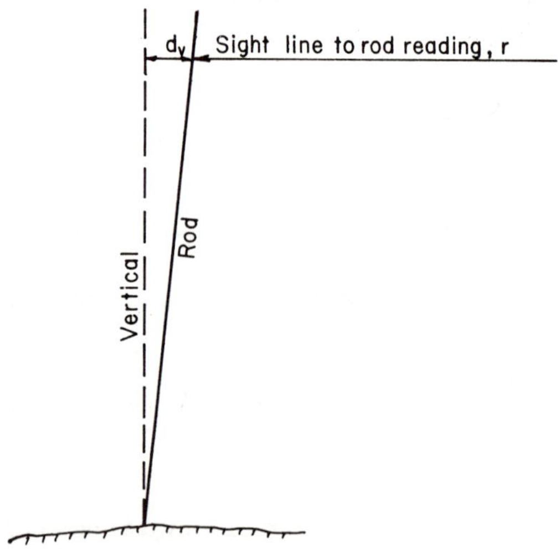

Figure 7.2 Rod Plumb Error

position of the rod, the error is directly proportional to the rod reading. This will be illustrated by an example later.

Centering bubble. On a dumpy, wye, or tilting level, the bubble axis and line of sight are parallel when in adjustment. The visible reference to assure that the line of sight is horizontal is the bubble centering marks. The two axes stay parallel and rotate together when the leveling screws are moved. Referring to Fig. 3.3 and Equ. 3.1 on bubble sensitivity, it can be visualized that a deviation, j, of the bubble axis also occurs along the line of sight if the bubble is off center by an amount, b, in a sight distance, D. If the bubble was off center by one division, Fig. 3.3 illustrates that the bubble axis, and thus, the line of sight rotates through the sensitivity angle, μ. Equation 3.1 will solve for the resulting rod reading error which is equal to j. Any fraction of a division can be estimated as an uncertainty in centering the bubble and from this, the uncertainty in a rod reading calculated as follows.

$$e_{b_1} = \pm f_d \mu D \tag{7.3}$$

where

e_{b_1} = the error in a rod reading at sight distance, D,

f_d = the fraction of a division which the bubble is estimated to be off center

μ = the bubble sensitivity in radians.

The bubble centering error is random in nature and random in total effect in the leveling process. The magnitude of the error depends upon the ability and care of the observer and, of course, upon the bubble sensitivity or radius and the sight distances. For a tilting level, an estimate can be made of the care in alignment of the bubble ends which can then be translated into an angular rotation in place of the product $f_d\mu$ in Equ. 7.3. Examples at the end of this section will illustrate the bubble centering error effects in differential leveling.

Reading the rod. This is purely a random error and is no different in nature than any reading error. It is primarily a function of the ability and care of the observer but is also influenced by the sight distance, clarity and value of rod markings, atmospheric conditions, telescope optics, and obstructions on the sight line. Its counterpart in theodolite work is the pointing error. Generally, the rod reading error is estimated as some fraction of the smallest division on a rod. For ordinary leveling using rods similar to those in Fig. 3.32, an estimate of ± 0.005 feet is commonly used.

Estimating sight distances. There is an error in a rod reading due to line of sight inclination which is compensated when backsight and foresight distances are equalized. Any error in keeping these distances equal will result in an error in the elevation difference if there is a line of sight inclination. The error is a function of the inclination, the care taken in keeping distances equal, and the length of the level circuit. Since the distances are usually only estimated or paced, there could be a fairly large uncertainty in the difference between backsight and foresight distances, which may or may not be significant in terms of elevation difference, depending on the magnitude of the line of sight error. Over a complete level circuit,

$$e_{SD} = \pm e_D E_{ir} \tag{7.4}$$

where

E_{ir} = inclination of line of sight in radians,

e_D = uncertainty in Σ BS distances - Σ FS distances.

Focussing. The effects of parallax were discussed and illustrated in Section 6-3. The same discussion applies to telescope focussing for leveling. If the eyepiece and objective lens are not both focussed in the reticle plane, the image of the horizontal cross-hair will move up and down slightly and thus an error in the rod reading can result if the eye of the observer is not directly along the line of sight axis. Parallax is easily detected by moving the eye slightly and can be easily corrected. Reference is made to Section 6-3 and Fig. 6.8 for a more complete discussion of parallax. Quantifying its effects in leveling is difficult. Any effects are best rendered negligible through proper focussing.

Natural Errors

Earth's curvature. Section 1.6 and Figs. 1.3 - 1.4 illustrate elevations and vertical distances. Since elevation differences are vertical distances between level surfaces, it would be ideal if the line of sight could somehow be caused to point along a curved line concentric to the vertical reference datum. But, since it cannot, individual rod readings will contain a positive error due to earth's curvature. These effects will be discussed simultaneously with effects due to atmospheric refraction.

Atmospheric refraction. Light rays are bent downward due to refraction effects of the atmosphere. A horizontal line of sight will thus be bent slightly in the direction of the earth curvature, but only about $^1/_7$ of the total effect of the curvature. Thus, individual rod readings will actually contain a negative error due to refraction alone.

The effects of earth's curvature and the combined effects of curvature and refraction are illustrated in Fig. 7.3. In this figure,

$$E_C = 0.667\ M^2 = 0.024\ F^2 \qquad (7.5)$$

$$E_{CR} = 0.574\ M^2 = 0.0206\ F^2 \quad (E_{CR} \text{ in feet}) \qquad (7.6a)$$

Also,

$$E_{CR} = 0.0675\ K_m^{\,2} \quad (E_{CR} \text{ in meters}) \qquad (7.6b)$$

where

M = the distance AB in miles ,

F = the distance AB in 1000's of feet,

K_m = the distance AB in Kilometers ,

E_C = curvature of earth effect on horizontal line of sight,

E_{CR} = combined curvature of earth and atmospheric refraction effect on horizontal line of sight,

E_{CR} = $^6/_7\ E_C$ (approximate).

If a telescope at A was pointed horizontally toward B and level rod held at B, the reading would be too high by E_{CR}. Therefore,

$$C_{CR} = -0.574\ M^2 \ (C_{CR} \text{ in feet}) \qquad (7.7a)$$

$$C_{CR} = -0.0206\ F^2 \quad (C_{CR} \text{ in feet}) \qquad (7.7b)$$

$$C_{CR} = -0.0675\ K_m^{\,2} \quad (C_{CR} \text{ in meters}) \qquad (7.7c)$$

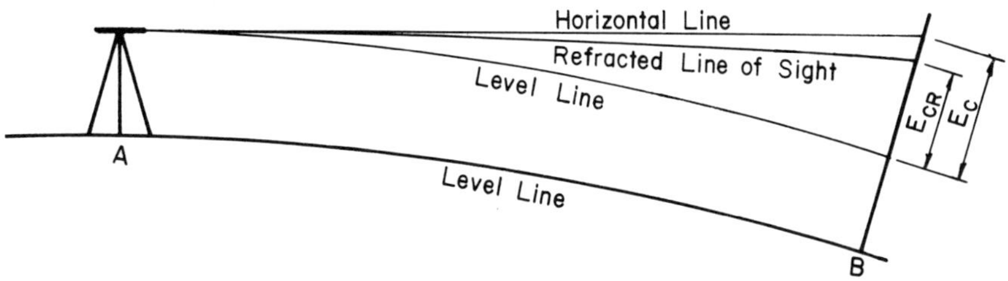

Figure 7.3 Earth Curvature and Atmospheric Refraction

is the correction to be applied to any rod reading. Although there are slight differences in the refraction, the differences are quite negligible as regards differential leveling. The errors cancel in the leveling procedures if sight distances are kept equal, but accumulate if there is a consistent difference in backsight and foresight distances, such as what might occur when leveling up or down a hill.

Settlement. If tripod legs are not pushed firmly into the ground or if the instrument is set up on marshy, thawing, or other unstable ground, the instrument could settle, resulting in a change in instrument height. It can be seen that if the settlement occurred in the time lapse between backsight and foresight readings, the latter reading would always contain a negative error. In the process of differential leveling this would accumulate since backsight readings are generally taken first with a time lapse occurring before the foresight reading is observed. The error could become significant in precise leveling when ground conditions are poor. It can be minimized by taking care to find stable ground for instrument set-ups. This may require chopping through icy ground if possible. It can also be compensated if, in alternate instrument set-ups, foresight readings are taken before backsight readings.

Settlement of the turning points under the weight of the level rod is also a possibility, but this is considered relatively minor if solid and stable points are selected.

Other natural errors. Section 6-4 included a discussion of wind, temperature changes, heat waves, and haze in reference to horizontal angle measurements. Much of that discussion and the precautions suggested also apply to leveling. Wind causes the level bubble to change centering and this must constantly be checked. It causes the compensator of an automatic level to vibrate, resulting in a rapidly oscillating reading which is nearly impossible to observe accurately on some instruments.

Just as in angle work, the best precaution to use as regards these natural errors is to avoid working in extreme conditions or to take whatever measures as are practical to reduce the effects.

Example 7-1. Given: An automatic level was tested for line of sight adjustment using the procedure based on Fig. 7.1. The readings were $a_1 = 5.21$ feet, $b_1 = 6.13$ feet, $a_2 = 4.88$ feet, $b_2 = 5.77$ feet. The sight distance, $D = 240$ feet. What is the inclination of the line of sight in feet/foot and in seconds, and what should be the reading at point B for adjustment?

Solution:

$$E_i = (a_2 - b_2) - (a_1 - b_1)$$

Substituting the given values,

$$E_i = \pm 0.03 \text{ ft.}$$

This is the error over sight distance 2D. The error per foot of sight distance is

$$E_{ir} = \frac{E_i}{2D} = \frac{0.03}{480} = 0.0000625 \text{ ft/ft (radians)}$$

or about 0.006 ft. per 100 ft. of sight distance. Converting this to seconds by multiplying by $\rho = 206{,}265$ sec/rad yields

$$E_i = 13 \text{ seconds.}$$

To adjust the instrument, the reading should be

$$a = (a_1 - b_1) + b_2 = 4.85 \text{ ft.}$$

That is, the line of sight must be moved down 0.03 feet from 4.88 to 4.85 feet while observing the rod held at point A on the last set-up.

Example 7-2. Given: A differential leveling circuit was run up a hill for a distance of 5 Kilometers. The starting BM elevation was 236.420 meters. After leveling back down the hill and adjusting an error of closure, the adjusted elevation for a BM set at the top of the hill was 398.160 meters. In the leveling process, the technicians sought to take advantage of the high rod and found they could use sight distances of 100 meters for the backsights going up the hill but foresights were restricted to about 50 meters each. Going down the hill, these sight lengths were reversed between backsights and foresights. After the circuit was run, the line of sight was tested and found to be inclined 10 seconds downward. What is the corrected elevation of the BM set on the hilltop?

Solution: It must be recognized that there are two error sources in this situation as described. They are the instrumental error in the line of sight and the natural error of curvature and refraction.

The length between turning points = 100 + 50 = 150 meters. Therefore, the total number of instrument set-ups in one direction is

$$n = \frac{5{,}000}{150} = 33.3 \quad \text{(use 34)}$$

Due to line of sight inclination, the error in a backsight going up the hill is

$$\begin{aligned} E_{BS_i} &= -10'' \text{ (0.00000485 rad/sec) 100 meters} \\ &= -0.00485 \text{ m,} \end{aligned}$$

and, in a foresight,

$$\begin{aligned} E_{FS_i} &= -10'' \text{ (0.00000485 rad/sec) 50 meters} \\ &= -0.00243 \text{ m.} \end{aligned}$$

There are 34 of each of these. Therefore,

$$E_{BS_{in}} = -0.00485\ (34) = -0.1649 \text{ m.}$$

$$E_{FS_{in}} = -0.00243\ (34) = -0.0825 \text{ m .}$$

The corrections to the sums of the readings would be

$$C_{BS_i} = +\ 0.165 \text{ m}$$

$$C_{FS_i} = +\ 0.083 \text{ m.}$$

Due to curvature and refraction, the error in a backsight going up the hill is

$$E_{CR} = 0.0675\ K_m^2 \tag{7.6b}$$

where

$$K_{BS} = 0.1 \text{ Kilometers}$$

from which

$$E_{CR\ BS} = 0.000675 \text{ meters.}$$

And, similarly for each foresight, $K_{FS} = 0.05$ Kilometers, and

$$E_{CRFS} = 0.00017 \text{ meters.}$$

The errors for the 34 set ups would be

$$E_{BSCRn} = 0.02295 \text{ meters} \qquad E_{FSCRn} = 0.00574 \text{ meters.}$$

The corrections to the sums of the readings would be

$$C_{BSCR} = -0.023 \text{ m} \qquad C_{FSCR} = -0.006 \text{ m.}$$

The ΣBS and the ΣFS must each be corrected by adding the sum of the applicable corrections. Thus, the true value for each backsight and foresight sum is:

$$\Sigma BS_T = \Sigma BS_R + \Sigma C_{BS}$$

and

$$\Sigma FS_T = \Sigma FS_R + \Sigma C_{FS}$$

or

$$\Sigma BS_T - \Sigma FS_T = (\Sigma BS_R + \Sigma C_{BS}) - (\Sigma FS_R + \Sigma C_{FS})$$

$$= (\Sigma BS_R - \Sigma FS_R) + (\Sigma C_{BS} - \Sigma C_{FS})$$

$$\Sigma C_{BS} = C_{BS_i} + C_{BSCR}$$

$$= +0.165 - 0.023 = +0.142 \text{ m}$$

$$\Sigma C_{FS} = C_{FS_i} + C_{FSCR}$$

$$= +0.083 - 0.006 = +0.077 \text{ m.}$$

$$(\Sigma BS_R - \Sigma FS_R) = 398.160 - 236.420 = 161.740 \text{ meters.}$$

$$\Delta Elev_T = \Sigma BS_T - \Sigma FS_T = 161.740 + 0.142 - 0.077$$

$$= 161.805 \text{ meters.}$$

The corrected elevation is

$$Elev_T = 236.420 + 161.805 = 398.225 \text{ meters.}$$

This problem illustrates the danger in not maintaining equal backsight and foresight distances. It is noted that these errors would not be revealed through a circuit closure on the same benchmark since the error signs would reverse when executing the circuit down the hill, thus cancelling.

Example 7-3. Given: The situation of Ex. 7-2 with the added condition that the 4-meter level rod was checked and found to have an error of +3 mm. in the 4-meter length. What additional correction must be applied to the adjusted elevation of the BM set on the hilltop?

Solution: An error of + 3 mm. in 4 meters requires a correction of

$$C_r = \frac{-3\text{ mm.}}{4\text{m}} = -0.75\text{ mm/m.}$$

assuming the error is uniformly distributed throughout the rod length. The actual readings are not known, nor is it necessary to correct each one or even the sums of backsights and foresights. When it is considered that the difference in these sums is all that is used to compute the elevation difference, the correction can be applied to this difference. It might be envisioned that, when leveling up the hill, there was just one large BS reading equal to the sum of individual readings and that the $\Sigma FS = 0$. Thus, the correction is

$$\begin{aligned} C_{BS} &= 161.74\text{ m }(-0.75\text{ mm/m.}) = -121.3\text{ mm} \\ &= -0.121\text{ meters} \end{aligned}$$

Thus, it is seen that a rod error which may seemingly be insignificant, can accumulate an appreciable error in a circuit where there is a significant elevation difference between the benchmarks. It is noted that the rod error would have caused no significant error if the terminal points had been at approximately the same elevation.

Example 7-4. Given: A 20-foot level rod which is one foot out of plumb at the 10-ft. mark. What would be the error in a reading at the 18 ft. mark? At the 3.6 ft. mark?

Solution:

$$E_v = \frac{d_v^2}{2r} \tag{7.2}$$

where

$$d_v = 1\text{ ft.}$$

and

$$r = 10\text{ ft.}$$

from which

$$E_v = 0.05\text{ ft.}$$

This is the error for a reading at the 10-ft. mark. Errors at other readings are directly proportional to the value of the reading.

$$E_{v18} = 1.8\ E_v = 0.09\text{ ft.}$$

$$E_{v3.6} = 0.36\ E_v = 0.02\text{ ft.}$$

A sketch of the configuration of the rod will aid in reasoning through the error situations demonstrated by this example. Among other things, this illustrates the importance of keeping the rod plumb for high rod readings.

Example 7-5. Given: An error estimate in each 80 meter sight distance of ± 5 meters in a level circuit consisting of 25 instrument set-ups, and the unadjusted line of sight of Ex. 7-1. What is the error in the difference in elevation observed due to sight distance and line of sight error?

Solution: It isn't known whether backsight distances were greater or less than foresight distances. This error is purely random, so it doesn't matter whether the line of sight inclination is plus or minus. The effect will be random. There are two sightings per set-up and 25 set-ups. Therefore there are 50 chances for error occurrence.

$$e_{SD} = \pm e_D E_{ir} \tag{7.4}$$

where

$$e_D = \pm 5\sqrt{50} = \pm 35 \text{ meters}$$

$$E_{ir} = 0.0000625 \text{ radians}$$

from which

$$e_{SD} = \pm 0.002 \text{ meters.}$$

Thus, it can be seen that for this line of sight inclination and circuit, the error due to estimating sight distances would be quite small even if the line of sight remained unadjusted, as long as attempts were made to equalize backsight and foresight distances. This particular circuit, it is noted, would have been 4 Kilometers long.

Example 7-6. Given: A metric rod with graduations of 1 centimeter and an automatic level having a compensator sensitivity of ± 0.25 seconds. It is estimated that the rod can be read to one-tenth of the least count or ± 0.1 cm. What is the error in one elevation difference if the backsight and foresight distances were each 100 meters?

Solution: There are two occurrences of each error in an instrument set-up because there are two sightings, the errors propagating as the law of addition or compensation. From Equ. 7.1

$$e_a = \pm 0.25'' \text{ (0.00000485 rad/sec) } 100\text{m.}\sqrt{2}$$

$$= \pm 0.00017 \text{ m}$$

$$e_r = \pm\, 0.001\sqrt{2} = \pm\, 0.0014 \text{ m}$$

$$e_{\Delta Elev} = \sqrt{e_a^2 + e_r^2} = \pm\, 0.0014 \text{ m}$$

Assuming that the estimate of the rod reading error is reasonable and would not be significantly lower, this example illustrates that the error due to compensator sensitivity is negligible in comparison to the personal rod error. This demonstrates the advantage of instrument precision and aids in fully comprehending the meaning of the term ''automatic'' level. The next example further develops this point.

Example 7-7. Given: A dumpy level used with the same rod and sight distances as Ex. 7-6. The level bubble has a sensitivity of 20 seconds per 2 mm graduation. It is assumed that the bubble is kept centered ± 0.5 divisions. What is the uncertainty in one elevation difference due to centering and rod reading error?

Solution: There are two occurrences for the elevation difference. From Equ. 7.3,

$$e_b = \pm f_d\, \mu\, D\, \sqrt{2}$$

where

$$f_d = \pm 0.5$$

$$\mu = 20 \text{ seconds } (0.00000485 \text{ rad/sec})$$

$$D = 100 \text{ m}$$

from which

$$e_b = \pm 0.007 \text{ m.}$$

The same rod reading error applies as calculated in Ex. 7-6, being $e_r = \pm 0.0014$.

$$e_{\Delta Elev.} = \sqrt{e_b^2 + e_r^2} = \pm\, 0.007 \text{ m.}$$

It is seen that with this instrument, the personal error in bubble centering, rather than rod reading, is the limiting factor on precision. Even if the bubble was kept centered ±0.25 division, it would still probably be the more significant error.

7-2 ANALYSIS OF TRIGONOMETRIC LEVELING MEASUREMENTS

Geometry of the Measurements

Trigonometric leveling employs the measurement of vertical angles and horizontal distances, both of which have been explained in Chapter 3. The field procedure and solution of the problem is relatively simple, especially when sight distances are short. When distances are long enough that the error due to earth's curvature and atmospheric refraction becomes significant, then this error must be corrected or compensated. Trigonometric leveling is often employed in terrain situations where differential leveling would be impossible or impractical, or over long distances in geodetic surveying where rapid but reasonably accurate results are required.

Figure 7.4 illustrates the general case where curvature and refraction are considered. In Fig. 7.4(a), a theodolite is at A and a rod is held vertically at B. The elevation difference between ground points A and B is,

$$\Delta Elev_{AB} = h.i._{A} + C_{CR} + V_{B} - r_{B}$$

where

$h.i._{A}$ = height of instrument above ground point A

C_{CR} = correction due to curvature and refraction

$V_{B} = L_{AB} \tan \gamma_{A}$

r_{B} = rod reading observed with the rod held at B

L_{AB} = horizontal distance between points A and B

γ_{A} = vertical angle to the point sighted on the rod (r_{B})

Curvature and refraction have been defined through Equ. 7.6.

In Fig. 7.4(b) the theodolite is at B and a rod is held vertically at A. The elevation difference between ground points B and A is,

$$\Delta Elev_{BA} = h.i._{B} + C_{CR} - V_{A} - r_{A} ,$$

where

$h.i._{B}$ = height of instrument above ground point B

$V_{A} = L_{AB} \tan \gamma_{B}$

r_{A} = rod reading observed with the rod held at A

γ_{B} = vertical angle to the reading r_{A} (note that γ_{B} is negative).

Actually, this equation is the same as that for $\Delta Elev_{AB}$ when the sign of the vertical angle is considered properly.

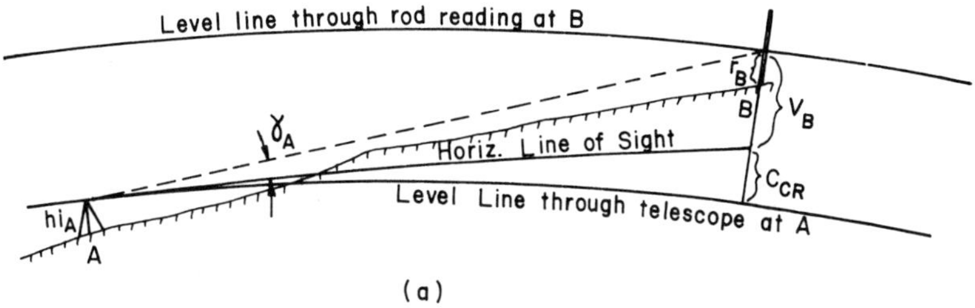

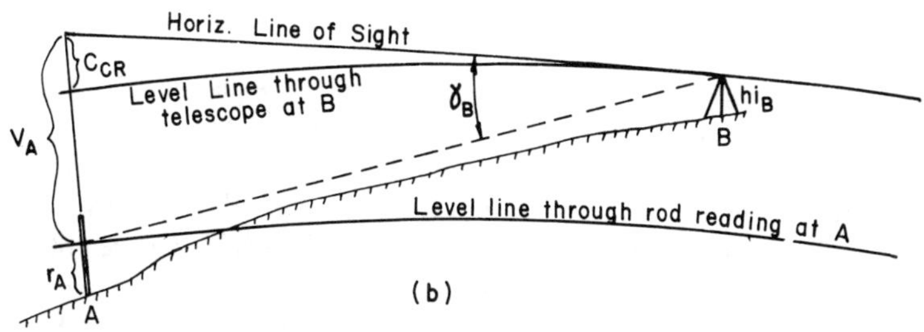

Figure 7.4 Trigonometric Leveling

Field Applications

There are several variations of the trigonometric leveling problem in the field. The simplest is where the vertical angle is measured from only one end of the line and the sight distance is short enough that curvature and refraction effects are considered negligible. In this case, the equation reduces to

$$\Delta \text{Elev}_{AB} = L \tan \gamma + \text{h.i.} - r \tag{7.8a}$$

Another case is where curvature and refraction are considered significant, the measurements are made from only one end of the line, and the vertical angle is positive.

$$\Delta \text{Elev}_{AB} = L \tan \gamma_A + \text{h.i.} - r + C_{CR} \tag{7.8b}$$

The third case is as above but where the vertical angle is negative. In this case, the tangent of the angle is negative.

$$\Delta \text{Elev}_{BA} = L \tan \gamma_B + \text{h.i.} - r + C_{CR}$$

or

$$\Delta Elev_{AB} = -(L \tan \gamma_B + h.i. - r + C_{CR}) \tag{7.8c}$$

Still another variation of the above, is to measure from both ends of the line. If this is done simultaneously or very close to the same time, high precision results since there is no *differential refraction*. This technique is called measuring *reciprocal vertical angles*. As will be shown below, the equal effects of curvature and refraction cancel when the average from both ends of the line is used. The sum of Equ. 7.8b and 7.8c, divided by two reduces to

$$\Delta Elev_{AB} = \tfrac{1}{2}[L(\tan \gamma_A - \tan \gamma_B) + h.i._A - h.i._B - r_A + r_B] \tag{7.8d}$$

It is noted that the sign of $\tan \gamma_B$ remains negative in this equation. If rod readings sighted are made equal to their respective h.i.'s, those values cancel. If terrain is very flat, both vertical angles may be observed as negative. This occurs because the higher point may not be greater in elevation than the curvature and refraction effect, C_{CR}. This should cause no confusion and the correct results are obtained if the algebraic signs are strictly observed in Equ. 7.8d.

Other variations of the trigonometric leveling problem exist where the theodolite may be set up between two points whose elevation difference is desired. This situation is similar to differential leveling except that vertical angles and the above equations are employed. If the instrument is halfway between the two points, the curvature and refraction effects cancel regardless of the elevation differences among the three points. The proof of this is left as an exercise for the reader.

A backsight can be observed on a nearby benchmark rather than measuring the h.i. if the ground point's elevation is not needed and an elevation is desired on the remote point. This is possibly the most precise technique since there is more error in measuring the h.i. at the instrument than reading a rod through a telescope. The full procedure is illustrated by a problem at the end of the chapter.

Errors in Trigonometric Leveling

All of the errors that were discussed for differential leveling also apply in trigonometric leveling. The line of sight error is compensated if vertical angles are observed as discussed in Section 3-3, reference being made to Fig. 3.17 where it is demonstrated how the errors are compensated in the direct and reverse process. Most systematic errors will not accumulate as in differential leveling since trigonometric leveling ordinarily involves only one or two instrument set-ups. The principal random errors are in measuring the vertical angles and the horizontal distance, measuring the h.i., reading the rod, and differential refraction. The errors due to the latter are ordinarily quite small.

Example 7-8. Given: A distance, AB, was measured using an electronic distance measuring instrument, the observed value being a slope distance of

2230.16 meters. A direct zenith angle observed to B with the theodolite at A was 90° 30′ 17″, and with the telescope inverted, it was 269° 29′ 23″. The h.i. = 1.80 cm which was also observed on the rod held at B. Reading and pointing error tests yielded a horizontal reading and pointing error of $E_{rp} = \pm 1.9''$ with 90% certainty. The EDM manufacturer states its precision as ±(1 cm. + D/500,000) where D is the distance. The h.i. and rod reading are assumed to be estimated to ± 0.5 cm. Assume this uncertainty and that of the EDM are 90% errors. Calculate the difference in elevation between points A and B and the uncertainty in this vertical distance.

Solution: The zenith angle must first be corrected for instrumental and reading errors:

	Observed	Correction	Corrected Value	
D	90° 30′ 17″	+10″	90° 30′ 27″	
R	269° 29′ 23″	+10″	269° 29′ 33″	
	359° 59′ 40″		359° 59′ 60″	check
	−359° 59′ 60″			
	E = −20″			

To be useful in the equations derived, the slope distance must be reduced to the horizontal. But, it is seen that this isn't necessary since, with slope distance the solution is

$$\Delta \text{ Elev}_{AB} = S \sin \gamma + \text{h.i.} - r + C_{CR} \tag{7.9}$$

where

$$S = \text{slope distance} = 2230.16 \text{ meters (given)}$$

$$\gamma = -0°30'27''$$

$$\text{h.i.} = r = 1.80 \text{ cm.}$$

$$C_{CR} = -0.0675\, K_m^2 = -0.0675\,(2.230)^2 \tag{7.7c}$$

from which

$$\Delta \text{ Elev}_{AB} = -19.753 + 1.80 - 1.80 + 0.336$$

$$= -19.417 \text{ meters.}$$

The uncertainties propagate according to the law of addition. However, the errors due to distance and angle must first be computed using the law of propagation of random errors in a product. It is assumed that the same reading and pointing error occurs in a vertical circle reading as in a horizontal circle reading. Using

$$V_B = S \sin \gamma$$

$$e_{V_B} = \sqrt{(S \cos \gamma)^2\, e_\gamma^2 + (\sin \gamma)^2\, e_s^2}$$

where

$$e_\gamma = \text{error in vertical angle} = \frac{\pm\, 1.9''}{\sqrt{2}} = \pm\, 1.34''$$

$$= \pm\, 0.00000651 \text{ rad}$$

$$e_m = \pm\, (1\text{cm} + \frac{223{,}016}{500{,}000}) = \pm 1.446 \text{ cm}$$

from which

$$e_{V_B} = \sqrt{2.108 + 0.00016} = \pm\, 1.45 \text{ cm}.$$

The student should note immediately that the error in elevation due to distance measurement is quite negligible in this problem. This error is, of course, a function of the vertical angle and the EDM precision, and, for steeper angles or less precise measuring techniques, this error source may not always be negligible. Now, the total propagated error is

$$E_{\Delta\ Elev} = \sqrt{e_{V_B}{}^2 + e_{h.i.}{}^2 + e_r^2 + e_{CR}{}^2}$$

where

$$e_{V_B} = \pm\, 1.45 \text{ cm}$$

$$e_{h.i.} = \pm\, 0.5 \text{ cm}$$

$$e_r = \pm\, 0.5 \text{ cm}$$

$$e_{CR} = \text{negligible}$$

from which

$$E_{\Delta\ Elev} = \pm\, 1.6 \text{ cm}.$$

It should be noted that it is mere coincidence that the value for e_m and for e_{V_B} are approximately the same. It is also noted that since the effect of e_s on the error in elevation is negligible, the solution of e_{V_B} reduces to

$$e_{V_B} = (S \cos \gamma)\, e_\gamma\ ,$$

and all of the squares and square root could have been omitted.

As another illustration of the approach introduced in Section 4-6, the above can be computed as follows:

$$V_B = S \sin \gamma$$

$$e_\gamma = (0°\ 30'\ 27'') + (1.34'') = 0°\ 30'\ 28.3''$$

$$V_B = 2230.16 \sin 0°\ 30'\ 27'' = 19.7535 \text{ m}$$

$$V_{B_e} = 2230.16 \sin 0° \; 30' \; 28.34'' = 19.7680 \text{ m}$$

$$e_{V_B} = V_B - V_{B_e} = 0.0145 \text{ m} = 1.45 \text{ cm}.$$

The same answer would result (with opposite sign) if the 1.34″ was subtracted from, rather than added to γ before re-computing the quantity.

The assumption that the error is negligible in the curvature and refraction constant may not be an accurate one. Since the magnitude of this uncertainty is difficult to estimate, reciprocal vertical angle measurements are suggested for greater precision.

The student is encouraged to compare the precision of trigonometric leveling with expected errors in differential leveling. Opportunity to do so will be afforded through the problems at the end of the chapter.

7-3 ANALYSIS OF ELEVATION DIFFERENCES BY STADIA

Geometry of the Measurements

Stadia theory was introduced in Section 5.2. Equations for determination of horizontal distances were derived, based on Fig. 5.4. A similar derivation can be done for vertical distance. The two components are generally considered together when employing stadia measurements in topographic and other surveys.

In a sense, stadia leveling is merely a variation of trigonometric leveling, the horizontal distance being determined by the stadia method. Figure 7.5 illustrates the situation for solution of the vertical distance with inclined stadia measurements. Reference is made to Section 5-2 for definition of terms and explanation of the measurements. In Fig. 7.5, it is seen that

$$V = S \sin \gamma$$

In Section 5-2, it was shown that

$$S = KI \cos \gamma + C_{SC}$$

Substituting this value for S into the first equation yields

$$V = KI \cos \gamma \sin \gamma + C_{SC} \sin \gamma \qquad (7.10a)$$

The value of $C_{SC} \sin \gamma$ is equal to zero when $C_{SC} = 0$, and it is quite small when γ is small, even for $C_{SC} = 1$. Another slightly simpler form of Equ. 7.10a can be derived from the trigonometric identity

$$2 \sin \gamma \cos \gamma = \sin 2\gamma \, .$$

Thus,

$$V = \tfrac{1}{2} KI \sin 2\gamma + C_{SC} \sin \gamma \qquad (7.10b)$$

The difference in elevation between the two ground points is

$$\Delta \text{ Elev} = V + \text{h.i.} - FS \tag{7.11}$$

Where

h.i. = instrument height above the ground

FS = rod reading.

If the FS is made to equal h.i., then the solution of Equ. 7.10b yields Δ Elev directly.

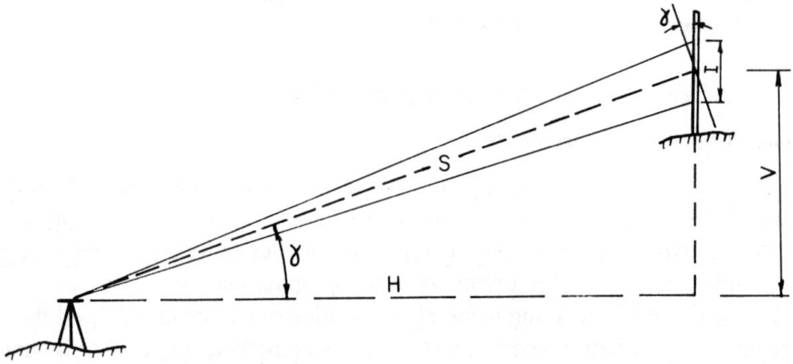

Figure 7.5 Stadia Leveling

It should be obvious that earth's curvature and atmospheric refraction are of little concern in stadia leveling since sight distances are generally only a few hundred feet and the effects at these typical distances are negligible when the stadia distance error is considered.

Field Applications

Leveling using stadia measurements employs a theodolite. If horizontal sights are possible, elevation differences are determined simply by differential leveling. If vertical angles must be measured between backsight and foresight, elevation differences are computed using Equ. 7.11.

One of the most common applications of Equ. 7.10 is in topographic surveying. With the instrument oriented along a known line and the HI determined, the horizontal and vertical position of any ground point within range can be determined by observing the direction, stadia distance, and vertical angle to the h.i. value. Points are then plotted horizontally using a protractor and scale and the elevation of each point computed using Equ. 7.10. This is the first stage in preparing a topographic map.

Errors in Stadia Leveling

Systematic errors in the rod and in the values for K and C_{SC} were discussed in Section 5-2. There it was stated that such errors are ordinarily neglected for horizontal distances, which is also true for vertical distances. Likewise, uncertainties in these values and in instrument centering are also negligible when the overall precision of the stadia method is considered. The important random errors are in measurement of I and γ.

Example 7-9. Given: The data of Ex. 5-10, with an h.i. equal to the mean of the upper and lower hair readings. What is the elevation difference between the instrument station and the rod station and the uncertainty in this elevation difference?

Solution: Assuming K = 100 and $C_{SC} = 0$

$$V = \tfrac{1}{2} KI \sin 2\gamma \tag{7.10b}$$

where

$$I = 2.71 \text{ ft.}$$

$$\gamma = -5° 16'$$

from which

$$V = 24.77 \text{ ft.}$$

$$\frac{\partial V}{\partial I} = \tfrac{1}{2} K \sin 2\gamma$$

$$\frac{\partial V}{\partial \gamma} = KI \cos 2\gamma$$

$$E_V = \sqrt{(\tfrac{1}{2} K \sin 2\gamma)^2 E_I^2 + (KI \cos 2\gamma)^2 E_\gamma^2}$$

where

$$E_I = \pm 0.005 \sqrt{2}$$

$$E_\gamma = (30/206{,}265) \text{ radians}$$

from which

$$E_V = \sqrt{0.0042 + 0.0015} = \pm 0.075 \text{ ft.}$$

There would also be an uncertainty in measuring the h.i. and subsequently pointing to this value on the rod, but these errors would be negligible with respect to the E_V calculated. In determining the significant figures of the elevation difference, it is noted that expression to tenths is appropriate. The value might be expressed as 24.8 ± 0.08 ft.

It can be seen that, just as in Ex. 5-10, the error resulting from the vertical angle uncertainty is negligible in comparison to the rod intercept error effect. The paragraph immediately following Ex. 5-10 pertains to the above example and should be reviewed.

Problems - Chapter 7

7.1 A level was tested and found to have a line of sight inclination of + 0.02 ft. per 100 ft. of sight distance. From a record of the sight distances, it was found that the backsight distances exceeded the foresight distances by 135 feet. What correction should be applied to the computed elevation difference between the two stations?

7.2 What inclination of the line of sight of a level would cancel the curvature and refraction error for sight distances of 100 meters?

7.3 A differential leveling circuit was run down a hill for a distance of 2.2 miles. Sight distances were estimated to be 250 ft. ± 25 feet by pacing. The tilting level, with a bubble radius of 50 meters, is assumed to be centered ± 1mm. by aligning the ends of the bubble through the optical microscope. The starting benchmark had a known elevation of 756.06 ft. The terminal benchmark was determined to have an elevation of 568.22 ft. before corrections were applied.The rod was found to be 0.015 ft. short at the 10-ft. mark. The rod reading error is assumed to be ± 0.005 ft. What is the true elevation of the terminal benchmark, corrected for systematic errors and the uncertainty in this elevation due to the random errors indicated?

7.4 Assuming there may be minor random errors in a differential leveling circuit due to nature and other sources which amount to perhaps ± 0.1 cm per instrument set-up, what would be the expected error of closure of a 1-Kilometer level loop using methods based on text Ex. 7-6?

7.5 What would be the expected closing error of a 1-Kilometer level loop using methods based on text Ex. 7-7 and assuming other miscellaneous random errors of ± 0.1 cm per instrument set-up?

7.6 A test of the sensitivity of a level bubble was made. With the rod at a distance of 250 ft., a reading of 4.980 ft. was observed with the bubble centered. The bubble was then moved off center 4 divisions and a reading of 5.053 ft. observed on the same rod. What is the sensitivity of the bubble?

7.7 Explain why the actual error of closure of text Ex. 7-2 does not affect the solution of the problem.

7.8 Text Ex. 5-11 was a computation of a distance and its uncertainty by the subtense bar method. During the same set-up, a vertical angle measurement of $4° 12' 24''$ was made to the center of the subtense bar. The uncertainty in this angle was ± 5 seconds. The height of the subtense bar to its sighting point was measured to be 1.75 m ± 0.005 m. The h.i. of the theodolite was 1.86 m ± 0.005 m. Compute the elevation difference between the two stations and the uncertainty in this value.

7.9 A horizontal distance of 426.8 ft. was taped to the center of the base of a flagpole. A mean vertical angle of 8° 15′ 30″ was calculated from measurements made with a 30-second direct reading vernier theodolite, using a direct and inverted observation. The distance is assumed to be measured using procedures for a relative distance error of 1/5000. A rod reading on a nearby bench mark was 5.20 ft. with the telescope horizontal. The BM elevation = 416.85 ft. A horizontal line of sight strikes the flagpole 2.92 ft. above its base. Where uncertainties aren't given use significant figures convention. Compute:
a) the elevation of the flagpole,
b) the height of the flagpole above its base,
c) the uncertainty in the elevation of the flagpole,
d) the uncertainty in the height of the flagpole above its base.

7.10 A horizontal distance of 2688.16 meters was measured to a distant station. A vertical angle of −2° 08′ 23″ was observed to the instrument's h.i. value on the rod. Compute the difference in elevation between the instrument and rod stations.

Additional Reading References - Chapter 7

1. Blachut, T.J., Chrzanowski, A., and Saastamoinen, J.H. *Urban Surveying and Mapping*. New York: Springer-Verlag, 1979, Chapter 4.

2. Brinker, Russell C. and Wolf, Paul R. *Elementary Surveying*. Sixth Ed. New York: Harper and Row, 1977, Chapter 5, 12.

3. Moffit, Francis H. and Bouchard, Harry. *Surveying*. Seventh Ed. New York: Harper and Row, 1981, Chapter 3, 19.

4. Schmidt, Milton O. and Rayner, William H. *Fundamentals of Surveying*. Second Ed. New York: D. Van Nostrand Co., 1978, Chapter 4.

CHAPTER 8

ANALYSIS OF INDIRECT MEASUREMENTS

8-1 GENERAL ANALYSIS OF INDIRECT MEASUREMENTS

Actually, the analyses of the last four chapters concerned indirect measurements since most of the measurements were a result of computations. This chapter contains examples and problems which are even more "indirect." Included are common surveying computational situations which involve the measured quantities of the previous chapters and their use to calculate additional desired quantitites such as departures and latitudes, coordinate positions, directional components, areas, volumes, and other quantities and dimensions. The emphasis is on how to compute the uncertainties in such indirect measurements.

It is hoped that after study of this chapter, the practicing surveyor will gain confidence in the use of coordinate systems and other methods of indirect measurement and seek to adopt all varieties of measurement systems as tools in everyday practice.

8-2 ANALYSIS OF TRAVERSE MEASUREMENTS

Basic Traverse Theory

Study of this section assumes some background in traverse computations. The basics of the calculations and field work are covered in several existing surveying texts. References are given at the end of the chapter.

A *traverse* is a series of connected lines whose lengths and directions are known. Traverses must start and end on points whose position is known in order to have geometric checks of the field work. Generally, a *closed traverse* is simply a closed polygon, but "closure" can be achieved by measuring between two points whose horizontal position with respect to each other is previously known. Closures in traversing are made for the same reasons as in leveling.

It should be obvious that a closed traverse has unique values for all distances and directions as they exist on the ground. Thus, it has a perfect closure, geometrically. But, because of errors in distance and angle measurements, and also earth curvature and meridian convergence effects for large areas, the figure will not "close" mathematically using the measurements. The tolerable error varies with specifications. The expected error is a function of errors in individual distance and angle measurements, as well as general traverse configuration, number of sides in the traverse, and the lengths of the traverse lines.

Traverses are computed by plane surveying methods which assume a plane surface with parallelism in meridians. Cartesian coordinates are thus employed, horizontal positions being defined by X and Y coordinates. Using the convention that X values increase to the east and Y values to the north, the difference in coordinate values between two ends of any traverse line, AB, are

$$\Delta X_{AB} = L_{AB} \sin Z_{AB} \tag{8.1a}$$

and

$$\Delta Y_{AB} = L_{AB} \cos Z_{AB} \tag{8.1b}$$

where

L_{AB} = horizontal distance of line AB ,

Z_{AB} = direction (azimuth) of line AB.

The values ΔX and ΔY are often referred to as *departure* and *latitude*, respectively. The sign of ΔX or ΔY is determined by the sign of sin Z and cos Z . Azimuth was defined in Section 1-6.

In a traverse ABC ..., the course BC has a departure (ΔX) and latitude (ΔY) as computed from its distance and direction. If the departure (or distance east-west) from A to C is desired, it is computed by adding $\Delta X_{AB} + \Delta X_{BC}$. The latitude (distance north-south) of AC is similarly computed. Thus, all traverse points are interrelated by their east-west and north-south components. A closure is computed merely by adding all ΔX's and all ΔY's as computed from the measured horizontal distances and directions. The uncertainty in the position of one point with respect to any other, and the expected error of closure can be derived using the principles contained in this paragraph and analysis of Equ. 8.1 using random error propagation.

Uncertainty in Position Between Adjacent Points

Applying the general formula for propagation of random errors to Equ. 8.1,

$$E_{\Delta X_{BA}} = \sqrt{(\sin Z_{AB})^2 E_{L_{AB}}^2 + (L_{AB} \cos Z_{AB})^2 E_{Z_{AB}}^2}\ , \tag{8.2a}$$

$$E_{\Delta Y_{BA}} = \sqrt{(\cos Z_{AB})^2 E_{L_{AB}}^2 + (L_{AB} \sin Z_{AB})^2 E_{Z_{AB}}^2}\ , \tag{8.2b}$$

where

$E_{\Delta X_{BA}}$ = uncertainty in position of B with respect to A, in X direction,

$E_{\Delta Y_{BA}}$ = uncertainty in position of B with respect to A, in Y direction,

$E_{L_{AB}}$ = uncertainty in length, AB,

$E_{Z_{AB}}$ = uncertainty in azimuth, AB.

Substituting Equ. 8.1a and 8.1b into 8.2a and 8.2b and slight rearranging results in

$$E_{\Delta X_{BA}} = \sqrt{(E_L \sin Z)^2_{AB} + (\Delta Y \cdot E_Z)^2_{AB}} \qquad (8.3a)$$

$$E_{\Delta Y_{BA}} = \sqrt{(E_L \cos Z)^2_{AB} + (\Delta X \cdot E_Z)^2_{AB}} \qquad (8.3b)$$

This can be illustrated as follows, for a traverse course in the first quadrant:

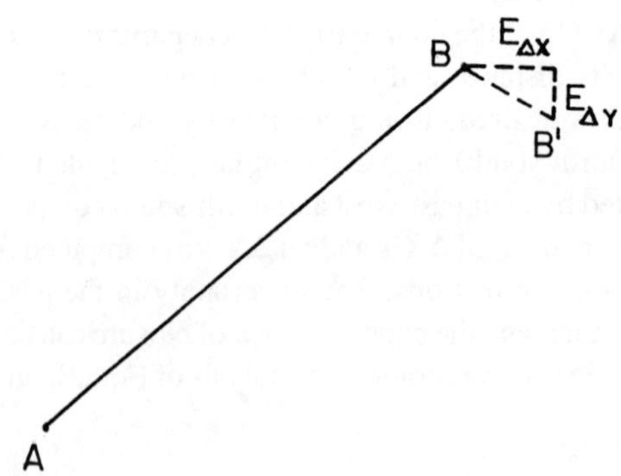

If one is not so familiar with the calculus, these equations can be derived from analysis of the component errors E_L and E_Z , illustrated as follows:

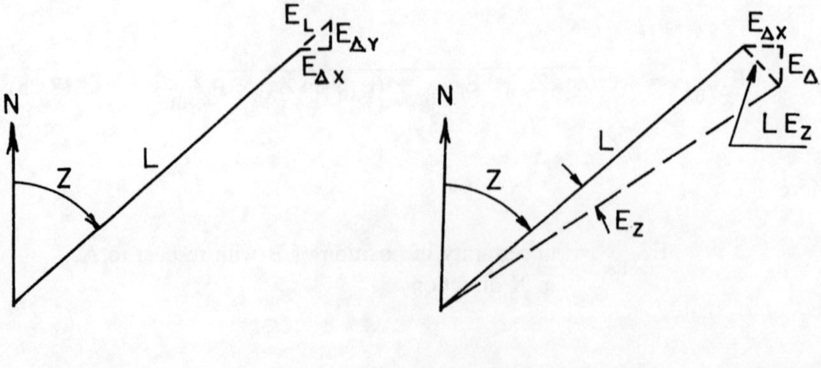

Due to E_L	Due to E_Z
$E_{\Delta X_L} = E_L \sin Z$	$E_{\Delta X_Z} = L\, E_Z \cos Z$
$E_{\Delta Y_L} = E_L \cos Z$	$E_{\Delta Y_Z} = L\, E_Z \sin Z$

Adding these respective components according to propagation of random errors in addition yields,

$$E_{\Delta X} = \sqrt{E_{\Delta X_L}^{\,2} + E_{\Delta X_Z}^{\,2}} = \sqrt{(E_L \sin Z)^2 + (L\, E_Z \cos Z)^2}$$

$$E_{\Delta Y} = \sqrt{E_{\Delta Y_L}^{\,2} + E_{\Delta Y_Z}^{\,2}} = \sqrt{(E_L \cos Z)^2 + (L\, E_Z \sin Z)^2}$$

which reduces to Equ. 8.2 and 8.3.

Example 8-1. Given: Point B has been set with respect to A along an azimuth of 40° and at 100.00 meters distance. The error in the azimuth is estimated to be ± 0.5′ and the error in distance as ± 2 cm. What is the error in position of B with respect to A?

Solution: This is a straightforward problem of substitution into Equ. 8.2 where

$$Z_{AB} = 40°$$
$$L_{AB} = 100.00 \text{ meters}$$
$$E_L = \pm\ 0.02 \text{ meters}$$
$$E_Z = \pm\ 0.5' = (30/206{,}265) \text{ radians.}$$

Equ. 8.2 yields

$$E_{\Delta X_{BA}} = \sqrt{0.0001653 + 0.0001241} = \pm\ 0.0170 \text{ m.}$$

$$E_{\Delta Y_{BA}} = \sqrt{0.0002347 + 0.0000874} = \pm\ 0.0179 \text{ m.}$$

The positional error is

$$E_{BA} = \sqrt{E_{\Delta X}^{2} + E_{\Delta Y}^{2}} = \pm\ 0.025 \text{ m.}$$

This example is simple enough that use of the derived equation 8.2 is unnecessary. Since the error in the position of B with respect to A is a direct function of the angular and distance uncertainties, these component errors, rather than the X and Y component errors, can be computed, then added for the positional error.

$$E_{BA_L} = E_L = \pm\ 0.02 \text{ m} \quad \text{(given)}$$

$$E_{BA_Z} = LE_Z = 100.00 \text{ m} \frac{30}{206{,}265} = \pm 0.0145 \text{ m}$$

$$E_{BA} = \sqrt{E_{BA_L}^{\ 2} + E_{BA_Z}^{\ 2}} = \pm\ 0.025 \text{ m.}$$

As the examples and situations become more complex, however, this type of solution becomes more cumbersome. Furthermore, all of the traverse error equations derived are easily programmed as part of the departure and latitude calculations making it easy to analyze the errors without seeking short-cuts.

It is worthwhile to note how traverse line configuration influences component errors ΔX and ΔY. For example, if a line is north-south, sin Z = 0 and cos Z = 1 causing the first term in Equ. 8.2a and the second term in Equ. 8.2b to be zero, resulting in $E_{\Delta X} = L \cdot E_Z$, $E_{\Delta Y} = E_L$. In other words, all of the error in ΔX is caused by the direction error and all of the error in ΔY by the length error. Similarly, if the line's direction is east-west, $E_{\Delta X} = E_L$ and $E_{\Delta Y} = L \cdot E_Z$ since sin Z = 1 and cos Z = 0 for Z = 90° or 270°. This should aid in determining the importance of controlling and analyzing errors for various traverse lines and traverse configurations. This point will be further discussed in Chapter 9.

Uncertainty in Position Between Non-Adjacent Points

In a survey, it may be of interest to know the uncertainty in one point with respect to another point, perhaps a control point. If the remote point whose positional error is desired is several traverse stations away from the control point, the solution is as follows.

Borrowing from the previous derivations,

$$E_{\Delta X_{CB}} = \sqrt{(\sin Z_{BC})^2 E_{L_{BC}}^{\ 2} + (L_{BC} \cos Z_{BC})^2 E_{Z_{BC}}^{\ 2}}$$

$$E_{\Delta Y_{CB}} = \sqrt{(\cos Z_{BC})^2 E_{L_{BC}}^{\ 2} + (L_{BC} \sin Z_{BC})^2 E_{Z_{BC}}^{\ 2}}$$

for the component errors in position of C with respect to B. If points B and C are traverse stations in traverse ABCD...., and the error in position of B with respect to A has been computed using Equ. 8.2, it is reasonable to expect that

$$E_{\Delta X_{CA}} = \sqrt{E_{\Delta X_{BA}}{}^2 + E_{\Delta X_{CB}}{}^2}$$

and

$$E_{\Delta Y_{CA}} = \sqrt{E_{\Delta Y_{BA}}{}^2 + E_{\Delta Y_{CB}}{}^2}$$

A more general equation for the positional uncertainty between any two non-adjacent traverse points is

$$E_{\Delta X} = \sqrt{\Sigma E_{\Delta x_i}{}^2} \qquad (8.4a)$$

$$E_{\Delta Y} = \sqrt{\Sigma E_{\Delta y_i}{}^2} \qquad (8.4b)$$

where

$\Sigma E_{\Delta x_i}{}^2$ = sum of the squares of the errors in ΔX for all traverse lines between the two points,

$\Sigma E_{\Delta y_i}{}^2$ = sum of the squares of the errors in ΔY for all traverse lines between the two points.

The square root of the sum of the squares of $E_{\Delta X}$ and $E_{\Delta Y}$ yields the positional error between the two points considered.

Uncertainty in Indirect Distance and Direction

This is actually a continuation of the previous analysis, except that uncertainty in the distance and direction between the two points is the result instead of ΔX and ΔY or positional error of one point with respect to the other. In a closed traverse whose coordinates (or departures and latitudes) are known, the distance and direction between any two non-adjacent points (for example, points A and C of traverse ABCD...A) is

$$L_{AC} = \sqrt{\Delta X_{AC}^2 + \Delta Y_{AC}^2} \qquad (8.5)$$

$$Z_{AC} = \tan^{-1} \frac{\Delta X_{AC}}{\Delta Y_{AC}} \qquad (8.6)$$

This is commonly known as the *inverse computation*.

The partial derivatives of the square root and tangent function are somewhat involved. Students should review basic calculus texts for the basic form of these derivatives. Taking derivatives of Equ. 8.5 yields

$$\frac{\partial L}{\partial \Delta X} = \frac{2\Delta X}{2\sqrt{\Delta X^2 + \Delta Y^2}} = \frac{\Delta X}{L} = \sin Z$$

Similarly,

$$\frac{\partial L}{\partial \Delta Y} = \frac{\Delta Y}{L} = \cos Z\,.$$

Derivatives of Equ. 8.6 are

$$\frac{\partial Z}{\partial \Delta X} = \frac{1/\Delta Y}{1+(\Delta X^2/\Delta Y^2)} = \frac{\Delta Y}{\Delta X^2 + \Delta Y^2} = \frac{\Delta Y}{L^2}$$

$$\frac{\partial Z}{\partial \Delta Y} = \frac{\Delta X/\Delta Y^2}{1+(\Delta X^2/\Delta Y^2)} = \frac{\Delta X}{\Delta X^2 + \Delta Y^2} = \frac{\Delta X}{L^2}$$

Substituting these partial derivatives into the general formula for propagation of random errors yields

$$E_L = \sqrt{(E_{\Delta X} \sin Z)^2 + (E_{\Delta Y} \cos Z)^2} \qquad 8.7)$$

$$E_Z = \frac{1}{L^2} \sqrt{(\Delta Y\ E_{\Delta X})^2 + (\Delta X\ \ E_{\Delta Y})^2} \qquad (8.8)$$

Example 8-2. Given: A closed traverse has been measured and computed. It is desired to know the distance and direction between two non-adjacent points and the uncertainties in these dimensions. Equation 8.4 has been used to compute the uncertainties in the relative ΔX and ΔY positions between the two points using error estimates for each individual directly measured distance and direction as computed. The result was $E_{\Delta X} = \pm\ 0.18$ feet and $E_{\Delta Y} = 0.08$ feet. An inverse computation using Equ. 8.5 and 8.6 yielded L = 1000.00 feet and Z = 40° 00′. What are the uncertainties in these dimensions and the error in position between the two points?

Solution:

$$\sin Z = 0.64279$$

$$\cos Z = 0.76604$$

$$\Delta X = L \sin Z = 642.8 \text{ feet}$$

$$\Delta Y = L \cos Z = 766.0 \text{ feet}$$

Substitution of the given and computed values into Equ. 8.7 and 8.8 yields

$$E_L = \pm 0.131 \text{ feet}$$

$$E_Z = 0.00014716 \text{ radians} \quad \pm 30.35 \text{ seconds}$$

The positional error caused by E_Z is

$$E_{P_Z} = L\, E_Z = 0.147 \text{ feet}$$

The error in position can be computed from

$$\sqrt{E_{P_Z}^{\,2} + E_L^{\,2}} = \sqrt{0.147^2 + 0.131^2} = 0.197 \text{ feet}$$

It is interesting to note that the positional error as computed from the given ΔX and ΔY component errors checks the above as computed from the length and direction component errors. That is,

$$\sqrt{E_{\Delta X}^{\,2} + E_{\Delta Y}^{\,2}} = \sqrt{0.18^2 + 0.08^2} = 0.197 \text{ feet.}$$

Expected Traverse Closure

The previous analyses provide a step toward determining the expected closure of a given traverse due to random errors in its individual distances and directions. In the traverse ABCD..., the uncertainty in ΔX and ΔY positions of any point with respect to the starting point is equal to the square root of the sum of the squares of all respective component ΔX and ΔY errors of each traverse line from the starting point to the terminal point. If this terminal point happens to be the starting point, as in closed traverse ABCD...A, the computation yields the expected closing error in ΔX and ΔY. From the theory used to derive Equ. 8.3 and 8.4,

$$E_{\Delta Xe} = \sqrt{\Sigma(E_{L_i} \sin Z_i)^2 + \Sigma\,(\Delta Y_i\, E_{Z_i})^2} \qquad (8.9a)$$

$$E_{\Delta Ye} = \sqrt{\Sigma(E_{L_i} \cos Z_i)^2 + \Sigma\,(\Delta X_i\, E_{Z_i})^2} \qquad (8.9b)$$

A further simplification of these equations can be made if the errors in all directions (E_Z) can be assumed to be equal and the errors in all distances (E_L) can be assumed to be equal. Thus,

$$E_{\Delta Xe} = \sqrt{E_L^2 \Sigma (\sin Z_i)^2 + E_Z^2 \Sigma \Delta Y_i^2} \tag{8.10a}$$

$$E_{\Delta Ye} = \sqrt{E_L^2 \Sigma (\cos Z_i)^2 + E_Z^2 \Sigma \Delta X_i^2} \tag{8.10b}$$

$$E_{ce} = \sqrt{E_{\Delta X_e}^2 + E_{\Delta Ye}^2} \tag{8.10c}$$

where

$$E_{ce} = \text{expected linear error of closure.}$$

The above equations provide a valuable tool in analyzing traverse closure. Once reasonable values for errors in all distances and directions (or angles) are estimated, the expected closure provides a means to check for careless work or mistakes. If the actual closing errors are greater than those computed by either Equ. 8.9 (unequal errors) or 8.10 (equal errors assumed), a problem in the field work would be suspected, causing concern for the overall accuracy of the points with respect to each other.

It may seem cumbersome to the practitioner to make calculations as above until it is realized that the equations are easily programmable and can be included in any traverse program by adding only a few computer statements and additional input of only the variables E_L and E_Z.

Example 8-3. Given: A four-sided travese was measured, using an assumed azimuth of one line. After computing the other three azimuths from measured interior angles, and computing each horizontal distance, values were as follows:

Course	Distance, Ft.	Azimuth
AB	1321.41	0° 12′ 20″
BC	2642.18	90° 22′ 10″
CD	1320.85	180° 18′ 30″
DA	2639.71	270° 21′ 20″

The estimated error in each distance is $E_L = \pm 0.05$ feet and the estimated error in each direction is $E_Z = \pm 5$ seconds. What is the expected closing error of this traverse?

Solution:

Course	sin Z	cos Z	ΔX	ΔY
AB	0.0035876	0.9999936	+4.741	+1321.409
BC	0.9999792	−0.0064480	+2642.125	−17.037
CD	−0.0053814	−0.9999855	−7.108	−1320.831
DA	−0.9999807	0.0062056	−2639.659	+16.381

Course	$(\sin Z)^2$	$(\cos Z)^2$	ΔX^2	ΔY^2
AB	0.00001287	0.99998720	22.477	1,746,121.75
BC	0.99995840	0.00004158	6,980,824.516	290.26
CD	0.00002896	0.99997100	50.524	1,744,594.53
DA	0.99996140	0.00003851	6,967,799.636	268.34
$E_{\Delta Xe}$ =	1.99996123	2.00003829	13,948,697.16	3,491,274.88

$$E_{\Delta XE} = \sqrt{(0.05)^2(1.99996) + (5 \times 0.000004848)^2\ (3{,}491{,}275)} \tag{8.10a}$$

$$= \sqrt{0.005 + 0.002} = \pm\ 0.084 \text{ ft.}$$

$$E_{\Delta Ye} = \sqrt{(0.05)^2(2.00\ldots) + (5 \times 0.000004848)^2\ (13{,}948{,}697)} \tag{8.10b}$$

$$= \sqrt{0.005 + 0.0082} = \pm\ 0.115 \text{ ft.}$$

$$E_{ce} = \sqrt{E_{\Delta Xe}^{\ 2} + E_{\Delta Ye}^{\ 2}} = 0.142 \text{ ft.} \tag{8.10c}$$

It is seen that the actual closures are

$$E_{\Delta X} = +0.099 \text{ ft.}$$

$$E_{\Delta Y} = -0.078 \text{ ft.}$$

yielding

$$E_{ca} = 0.126 \text{ ft.}$$

Although the actual closure in ΔX was slightly higher than expected, the linear closure was less than expected; therefore, the survey would be considered satisfactory. The student should realize the inherent weakness in assuming that errors in all distances and directions are equal. For traverses with significant differences in traverse line lengths, this may not be a good assumption, and therefore solutions should generally employ Equations 8.9 with 8.10c.

The alternate solution for this problem avoids the propagation of ΔX and ΔY component errors and instead propagates the errors due to length and direction directly. This may seem more straightforward and is as follows:

For each length, the distance error is given as $\pm$ 0.05 ft. These can be added in one step

$$E_{ce_L} = \pm 0.05\sqrt{4} = \pm 0.100 \text{ ft. } (n = 4 \text{ sides, equal errors})$$

The error, in radians, due to direction is

$$\pm \frac{5}{206{,}265} = \pm\ 0.00002424$$

Multiplying this by each course length yields its respective error E_{ce_Z} which is the effect on closure due to the error in that particular direction.

Course	E_{ce_Z}
AB	0.032
BC	0.064
CD	0.032
DA	0.064

Adding these errors using the usual propagation in addition theory yields

$$E_{ce_Z} = \sqrt{0.032^2 + 0.064^2 + 0.032^2 + 0.064^2}$$

$$= \pm 0.101 \text{ ft.}$$

and thus,

$$E_{ce} = \sqrt{E_{ce_L}{}^2 + E_{ce_Z}{}^2} = \sqrt{0.100^2 + 0.101^2}$$

$$= \pm 0.142 \text{ ft.}$$

8-3 ANALYSIS OF TRIANGULATION MEASUREMENTS

The complexities of triangulation measurements involving the quadrilateral and other control extension systems are not within the scope of this text, although uncertainties in such measurements are possible to compute using propagation methods introduced herein.

One of the simplest indirect measurement techniques available to the surveyor is intersecting for the position of a point or for determining the length of a line. Intersection is often used when a point is inaccessible or when terrain cannot be crossed for distance measurement. Even though EDM can sometimes be used for this problem, intersection is still used for many field situations. The geometry of intersection is illustrated as follows:

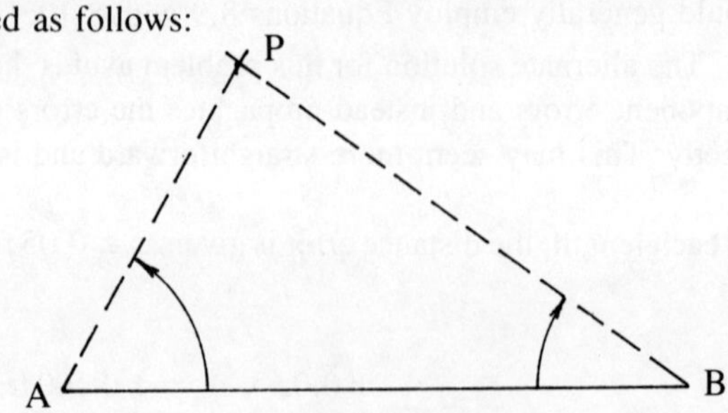

A theodolite is set up over points A and B and horizontal angles measured to inaccessible point P. Horizontal distance AB is determined. The triangle is solved by the law of sines.

$$\psi = 180° - \alpha - \beta$$

$$AP = \sin \beta \frac{AB}{\sin \psi}$$

$$BP = \sin \alpha \frac{AB}{\sin \psi}$$

If the coordinate of either point A or B and the direction AB is known, coordinates of P can be determined. For example,

$$Z_{AP} = Z_{AB} - \alpha$$
$$\Delta X_{AP} = AP \sin Z_{AP}$$
$$\Delta Y_{AP} = AP \cos Z_{AP}$$

If the uncertainty in measurement of AB, α, β, and Z_{AB} are known or assumed, it can be seen that subsequent calculation of the uncertainty of either AP or BP and the position of point P is merely another application of the general formula for propagation of random errors. The following example illustrates this error analysis.

Example 8-4. Given: The distance across a busy highway is desired. Measurements have been made along a right-of-way line on the south side of the highway. Angle measurements have been made from two points on the right-of-way to a control point on the opposite side, all as shown in the sketch. Uncertainties in all measurements are also shown. What is the highway right-of-way width between the two control points AD, and the uncertainty in this dimension?

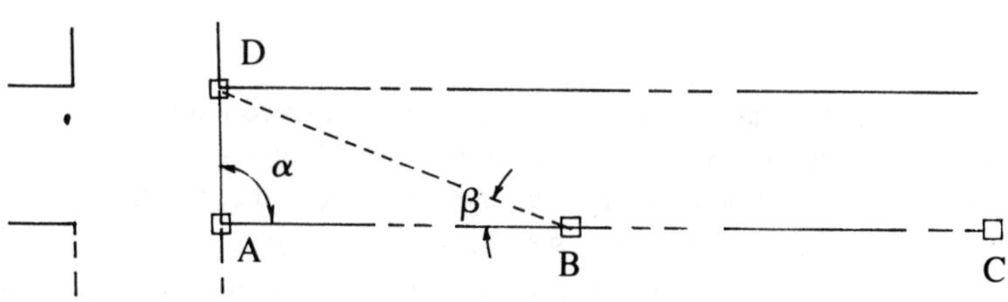

$$AB = 271.41 \pm 0.02 \text{ ft.}$$
$$\alpha = 90° \, 01' \, 15'' \pm 15''$$
$$\beta = 20° \, 15' \, 30'' \pm 15''$$

Solution:

$$\psi = 180° - (\alpha+\beta) = 69° \; 43' \; 15''$$

$$AD = AB \frac{\sin \beta}{\sin \psi} = 100.187 \text{ ft.}$$

$$e_\psi = \sqrt{e_\alpha^2 + e_\beta^2} = \pm 15 \sqrt{2}''$$

$$\frac{\partial \, AD}{\partial \, AB} = \frac{\sin \beta}{\sin \psi}$$

$$\frac{\partial \, AD}{\partial \beta} = \frac{AB \cos \beta}{\sin \psi}$$

$$\frac{\partial AD}{\partial \psi} = AB \sin \beta \csc \psi \cot \psi = \frac{AB \sin \beta}{\sin \psi \tan \psi}$$

$$e_{AD} = \sqrt{\left(\frac{\partial AD}{\partial AB}\right)^2 e_{AB}^{\;2} + \left(\frac{\partial AD}{\partial \beta}\right)^2 e_\beta^{\;2} + \left(\frac{\partial AD}{\partial \psi}\right)^2 e_\psi^{\;2}}$$

Substituting the measurements and their errors into this equation and the equations for evaluating the partial derivatives results in

$$e_{AD} = \sqrt{(0.00738)^2 + (0.01974)^2 + (0.003806)^2}$$

$$= \pm 0.021 \text{ ft.}$$

This example illustrates how a distance can be measured indirectly with an uncertainty approximating what may have been possible using either a tape or EDM. The reader should note the uncertainties assumed and compare the result with what may have been possible using a theodolite with precision higher than ± 15 seconds. Also, if the error sources are investigated, it is seen that the measurement of angle B contributed most to the total propagated error. Further investigation of the geometry of intersection would reveal that triangles with small angles do not give the best intersected values because of the rate of change in some of the trigonometric functions. This is why it is said that in triangulation, very small or very large angles should be avoided. The proof of this is left as an exercise for the interested student.

It should also be observed that if the coordinates of points A and B were known in Ex. 8-4, the coordinates of point D could be computed, along with the uncertainty in those X and Y values. This would involve taking derivatives of Z_{AD}, ΔX_{AD}, and ΔY_{AD}.

8-4 ANALYSIS OF AREA MEASUREMENTS

Areas are computed in several ways in surveying, the method of coordinates and double meridian distances being the most common two methods for traverses. The latter method (DMD's) can be used to illustrate the approach for estimating the error in a computed traverse area. The DMD method states that the area is one-half the sum of each DMD multiplied by the respective latitude.

$$\text{Area} = \tfrac{1}{2} \Sigma \, (\text{DMD}_i \times \Delta Y_i)$$

where a DMD is equal to the DMD of the preceding course plus the departure of that course plus the departure of the course itself. The DMD of the first course is equal to its departure. This method should be studied in available references if the student is unfamiliar with it. For a three-sided traverse ABC,

$$\begin{aligned} \text{Area} = \tfrac{1}{2}[&(\Delta X_{AB}\,\Delta Y_{AB}) + (\Delta X_{AB} + \Delta X_{AB} + \Delta X_{BC})\,(\Delta Y_{BC}) \\ &+ (\Delta X_{AB} + \Delta X_{AB} + \Delta X_{BC} + \Delta X_{BC} + \Delta X_{CA})(\Delta Y_{CA})] \end{aligned}$$

which, when expanded is

$$\text{Area} = \tfrac{1}{2}\Delta X_{AB}\Delta Y_{AB} + \Delta X_{AB}\Delta Y_{BC} + \tfrac{1}{2}\Delta X_{BC}\Delta Y_{BC} + \Delta X_{AB}\Delta Y_{CA} + \Delta X_{BC}\Delta Y_{CA} + \tfrac{1}{2}\Delta X_{CA}\Delta Y_{CA}$$

$$\frac{\partial\ \text{Area}}{\partial \Delta X_{AB}} = \tfrac{1}{2}\Delta Y_{AB} + \Delta Y_{BC} + \Delta Y_{CA}$$

$$\frac{\partial\ \text{Area}}{\partial \Delta X_{BC}} = \tfrac{1}{2}\Delta Y_{BC} + \Delta Y_{CA}$$

$$\frac{\partial\ \text{Area}}{\partial \Delta X_{CA}} = \tfrac{1}{2}\Delta Y_{CA}$$

$$\frac{\partial\ \text{Area}}{\partial \Delta Y_{AB}} = \tfrac{1}{2}\Delta X_{AB}$$

$$\frac{\partial\ \text{Area}}{\partial \Delta Y_{BC}} = \Delta X_{AB} + \tfrac{1}{2}\Delta X_{BC}$$

$$\frac{\partial\ \text{Area}}{\partial \Delta Y_{CA}} = \Delta X_{AB} + \Delta X_{BC} + \tfrac{1}{2}\Delta X_{CA}$$

When these partials are substituted into the general equation for propagation of random errors, the uncertainty in the area of this triangle would result. For the general situation of *n* sides, partial derivatives would need to be taken with respect to each ΔX and ΔY. But, a sequence is noted in the above equations. Once this is observed, the partial derivatives for each course can be written without expanding each equation for area. The sequence should become obvious upon study of the above derivatives. The solution for the error in traverse areas is rather lengthy but can be programmed on a computer.

8-5 ANALYSIS OF MISCELLANEOUS INDIRECT MEASUREMENTS

The reader may, by this time, have been stimulated to think of several instances of application of the theory of error propagation to indirect measurements. Some fairly common applications have been detailed. Others include miscellaneous areas, volumes, position by resection, height of inaccessible point, astronomic azimuths, elements of a circular curve, position of points by intersection of curves or lines, and indirectly computed angles. These applications will be illustrated through the problems at the end of the chapter.

Problems - Chapter 8

8.1 What is the departure and latitude of a traverse course with length 4626.81 ft. and azimuth 286° 12′ 56″?

8.2 Referring to Problem 8.1, if the uncertainty in the length is ±0.5 ft. and in direction, ±5″, what is:

a) the uncertainty in the departure and latitude?

b) the uncertainty in relative position between the two ends of the line?

8.3 Is there "consistency" between the length and direction measurements of Problem 8.1 — that is, are positional errors caused about as much by distance as direction errors?

Work problems 8.4 through 8.7 using the following data for traverse ABCDEFA

Line	Azimuth, Z	Horizontal Length, L	e_Z, ″	e_L, ft.
AB	89° 56′ 31″	2643.89	± 5″	± 0.25
BC	359° 57′ 43″	2645.66	± 5″	± 0.25
CD	270° 00′ 38″	1320.98	± 10″	± 0.15
DE	179° 56′ 04″	1322.34	± 10″	± 0.15
EF	270° 01′ 16″	1321.45	± 10″	± 0.15
FA	180° 02′ 38″	1326.90	± 10″	± 0.15

8.4 Compute the departures, latitudes, actual closure in departure and latitude, linear error of closure, and relative error of closure of the traverse.

8.5 Compute the expected error of closure in departure and latitude, and the expected linear error of closure using the errors given.

8.6 Compute the length and direction of line CF and the uncertainty in these values.

8.7 Compute the area enclosed by the lines CD, DE, EC and the uncertainty in this area.

8.8 From two control stations, A and B, the coordinates of a radio tower, C, are to be determined by intersection. The data are as follows:

$X_A = 516{,}815.21$ $\qquad Y_A = 225{,}618.45$

$X_B = 516{,}995.30$ $\qquad Y_B = 230{,}775.16$

$\measuredangle$ BAC = 39° 12′ 26″± 3″ $\measuredangle$ CBA = 57° 30′ 40″± 3″

Assuming the two control points are without error, compute the X and Y coordinates of point C and the uncertainty in these coordinates.

8.9 A land survey measurement had to be made from a traverse offset line because thick brush and a fence on line prevented accurate direct measurement. The offset line data are shown.

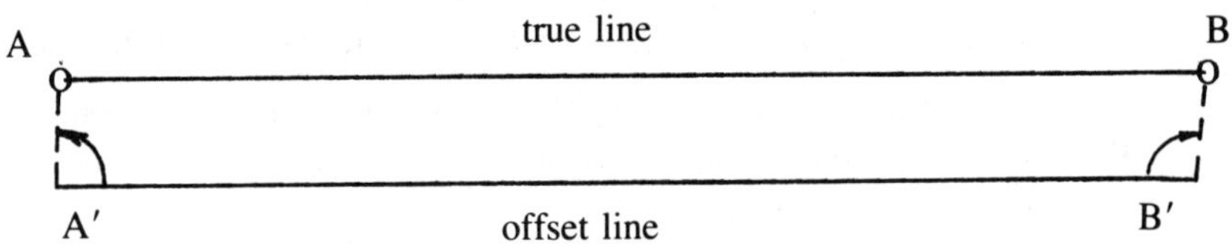

$$\begin{aligned} AA' &= 20.32 \text{ ft.} \pm 0.02 \text{ ft.} \\ \measuredangle A' &= 86°12.5' \pm 0.5' \\ A'B' &= 475.86 \text{ ft.} \pm 0.05 \text{ ft.} \\ Z_{A'B'} &= \text{N } 89° \ 12.0' \text{ E} \pm 1.0' \\ BB' &= 21.16 \text{ ft.} \pm 0.02 \text{ ft.} \\ \measuredangle B' &= 102° \ 17.5' \pm 0.5' \end{aligned}$$

What is the length and direction of line AB and the uncertainty in these values?

8.10 The area of a pie-shaped wedge of land is to be computed, one side being a segment of a circle with radius 200.00 ± 0.01 meters and central angle 36° 14′ 10″±5″. Compute the area in hectares and the uncertainty in this value.

8.11 A cyclindrical hole has been drilled for a caisson. The hole is 36 inches in diameter ± 3 inches, and 40 ft. deep ± 1ft. What is the volume of dirt removed from the hole (in its original, natural condition) and the uncertainty in this volume, in cubic yards?

8.12 The elevation of the top of a high-rise apartment building is desired. It is unsafe and inconvenient to make direct measurements, so a system of triangulation, combined with trigonometric leveling is employed. Field data are as follows:

$$\begin{aligned} AB &= 289.024 \pm 0.05 \text{ ft.} \\ \measuredangle A &= 62° \ 41' \ 45'' \pm 10'' \\ \measuredangle B &= 104° \ 19' \ 18'' \pm 10'' \end{aligned}$$

Vertical $\measuredangle$ at A to building top

$$= 10° \ 31' \ 56'' \pm 10''$$

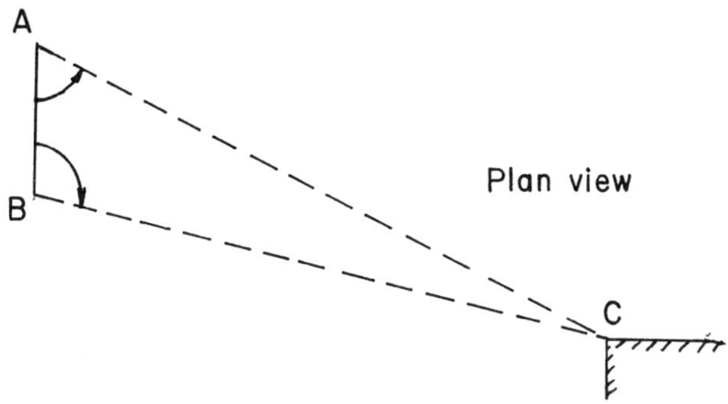

BS on BM from A = 2.42 ± 0.01 ft. (telescope horizontal)
Elev BM = 729.383 ± 0.02 ft.

What is the elevation of the top of the building and the uncertainty in this elevation?

8.13 An astronomic azimuth by Polaris observation can be computed by equation $Z = p\frac{\sin t}{\cos h}$ where Z = azimuth, p = polar distance (minutes), t = hour angle (degrees), h = altitude (vertical angle, degrees). From judgments of uncertainties in interpolating ephemeris tables, and observing time, angle, etc., it is estimated that uncertainties are: p, ± 0.005 minutes arc; t, ± 1 minute arc; h, ± 30 seconds arc. Given that p = 49.50 minutes, t = 118° 12′ 50 , and h = 40° 12′ 45″, compute the uncertainty in Z for one pointing to Polaris.

8.14 Using four pointings to Polaris as a backsight as in Problem 8.13, with four pointings to a survey line target as a foresight, and given that the combined reading and pointing error of the theodolite is ± 2.0″ to a well-defined target, what is the uncertainty in azimuth to the ground target from error sources mentioned?

8.15 Compute the area and the uncertainty in the area of a triangle with two sides and the included angle measured as follows:

b = 250.00 m ± 3 cm. c = 180.00 m ± 2.5 cm.
∢ A = 40° 00′ ±15″

8.16 Compute the value of the angle B (opposite side b) in Problem 8.15, and the uncertainty in this value.

8.17 National Map Standards dictate that 90% of the well defined map points are to be located within 1/50 inch on the finished map. What is the 90% error in a distance scaled from a 1:24,000 map between two well defined map points if the scaling error itself is ± 0.01 inch?

8.18 Given that directions in a survey have been determined ± 10 seconds, in what decimal place does the cosine of the following azimuth values start to change?

0° 26′ 18″	135° 40′ 36″
36° 49′ 42″	241° 32′ 50″

8.19 The tangent distance of a horizontal circular curve is computed by $T = R \tan \Delta/2$ where R is the radius of the circle and Δ is the central angle subtended between the point of curvature and point of tangency. Given that the radius and central angle have been measured as R = 150.02 ± 0.03 ft. and $\Delta = 38° \ 12.8' \pm 0.05'$, what is the tangent distance and its uncertainty?

8.20 The length of a chord between two points on a circular arc is computed as $c = 2R \sin \theta/2$ where R is the radius of the circle and θ is the angle subtended by the two points on the arc defining the chord. If the radius is assumed to be without error, what is the θ value and its uncertainty for a 1000 ft. radius circular curve with a chord measured as 361.85 ± 0.04 ft.?

8.21 A marathon road race is measured by laying off a 10 mile loop (which will be run twice) and an additional ''out-and-back'' portion of 3 miles plus 192.5 yards to complete the regulation distance. A bicycle wheel counter was calibrated over a known distance of 2415.85 ± 0.02 ft. Wheel revolutions of 7236, 7238, 7237, 7236 were recorded for each of four trips over the calibration course. Assuming no errors due to tire pressure changes or other sources, and that the 10 mile loop and 3+ mile segment were measured only once, what is the error in establishing the full marathon distance (26 miles, 385 yards) due only to random errors in calibration?

8.22 Assume that an auto odometer can be accurately calibrated using interstate highway mile posts over long distances arriving at a correction factor with ± 0.002 mi/mi precision. Assume also that one can read initial and final values to ±0.02 miles from the odometer. What would be the uncertainty in the distance of Problem 8.21 using a calibrated auto odometer?

Additional Reading References - Chapter 8

1. Brinker, Russell C. and Wolf, Paul R. *Elementary Surveying*. Sixth Ed. New York: Harper and Row, 1977, Chapter 10.
2. Moffitt, Francis H. and Bouchard, Harry. *Surveying*. Seventh Ed. New York: Harper and Row, 1981, Art 4-7, Chapter 8.
3. Schmidt, Milton O. and Rayner, William H. *Fundamentals of Surveying*. Second Ed. New York: D. Van Nostrand Co., 1978, Chapter 7.

Section C

CONTROLLING MEASUREMENTS

CHAPTER 9

THE SURVEYOR AS EXPERT MEASURER

9-1 THE ROLE OF THE SURVEYOR

The problem of measurement generally reduces to a five step procedure:

1. Selection of the method or technique to use for a specific measuring task,
2. derivation and adoption of specifications and standards,
3. execution of the measurements and recording of data,
4. employment of various checks and controls during execution,
5. analysis and reduction of data.

The choice of equipment, number of repetitions, level of care taken in making readings and handling equipment, and manner of making independent checks, all affect the time required for the survey, which in turn affects costs. The trade-offs are always between time-costs plus other expenses vs. accuracy achieved. A less accurate survey can, in general, be done much cheaper since less time and less refined instruments are required. Likewise, if the tolerances are to be very strict, then more expensive equipment and possibly more time and care are required. The professional challenge to a surveyor is often in choosing the methods that satisfy the accuracy requirements at the least possible cost.

A well-known surveying educator has stated:

> The undisputable fact that no measurement is exact, must be acknowledged, for exactness is impossible in the physical world. The man who regards his measurements as being "right on the button" is only deceiving himself. In fact, he may be very shocked to discover how inaccurate his measurements really are! Every measurement has a certain degree of uncertainty. Indeed, the mark of a good surveyor may well be his ability to recognize and control the uncertainty of his measurements. (Ref. 6 pp. 423-28.)

Some words of importance in the above introductory comments are.."selection"..."derivation"..."adoption"..."execution"..."checks"..."analysis"..."choice"..."care"..., and above all "*control*". All of these words imply that a surveyor must be able to think and make decisions. All of these words connote the role of a professional in any field, and being in control generally covers nearly all other considerations. An automobile out-of-control has an inexperienced or reckless operator steering it. A professional task or project can similarly get out of control unless the engineer (used in a general sense) plans, thinks, supervises, and makes decisions. Few surveying jobs are alike. Therefore, they cannot all be executed the same way. The key to success is to understand the theory and the execution. Analytical thinking is nearly always involved.

The public has a right to expect that a professionally licensed surveyor will be in control at all times and will give them appropriate accuracy at the least possible cost. They have this right to demand professional performance since licensing has the sole purpose of protecting the public from incompetent practitioners. There are no other listings in the telephone directories or federal job classifications which imply expertise in land measurement. The public seeks the services of a surveyor *primarily* because they *expect* the surveyor to be an *expert* in the science and art of measurement, and the courts back them up in this expectation. The well known text, *Land Surveying* by F. Hodgman contains several citations of court cases which reveal that surveyors, according to law, have no recognized expertise *other* than measurement. For example:

> Surveyors have no more authority than other men to determine boundaries of their own motion. (Cronin V. Gore, 38 Mich. 381, p. 378 Hodgman).

> The law recognizes surveyors as useful assistants in...measurement and calculation...But the determination of facts belongs exclusively to the courts and juries. (Stewart V. Carleton, 31 Mich. 273, p. 378 Hodgman).

And, in another well known text, *Evidence and Procedures for Boundary Location* by Brown and Eldridge, the role and legal function of a surveyor is firmly explained, as follows:

> The surveyor is an expert measurer. In court cases his value as an expert is primarily in his testimony as to distances and directions. (Brown and Eldridge, p. 441).

Brown and Eldridge continue with:

> ...The surveyor should never be so foolish as to represent his measurement as being exact and without error, for a wise attorney, upon cross examination, will quickly discredit such testimony. A frank statement of expected error and the sources of errors must be presented... (Brown and Eldridge, p. 441).

The very first role of a surveyor is one of expert measurer. All else is secondary. All of the best business and legal background will not assure that accuracy is being achieved at a fair price to the consumer and for the benefit of future generations who must utilize the data and the ground survey information established.

Once this primary role is accepted, the surveyor can proceed to refine techniques and practices, always working toward being a true expert, in full control of measurements. The main part of this text provides one major source of background toward this end. These next few sections add a few refinements to further develop the reader into a true expert in the chosen field of surveying.

9-2 CONTROLLING UNCERTAINTY IN MEASUREMENTS

To be in control of measurement reduces primarily to being in control of uncertainty. How uncertain and how confident of the uncertainty is the question. Anyone can be taught to make readings or turn knobs on instruments, but going that step beyond the size into the dimension of evaluating uncertainty, says Gracie (see earlier quote), is the true mark of a surveyor. But, what does it mean to be in control, particularly when the true value cannot be determined anyway? In the final analysis, if the surveyor can, with confidence state the numerical range within which each measurement could be and the level of certainty (percent probability) of this range then control is evident. In other words, random error theory, based either on judgment or statistical analysis must be employed in order to estimate all of importance concerning any particular measurement.

Control also means to be able to wisely select methods or equipment which either increase or decrease precision and accuracy according to specifications and/or costs. Again, to do this properly, application of random error theory is important. This may mean complete design of measurement specifications. It may simply mean that the surveyor notices the results of a certain method repeatedly yield much better accuracy than is required and decides to modify the system. The modification may mean substituting a less expensive instrument, resulting in a less costly job and opportunity to use the more precise instrument elsewhere.

Being in control of uncertainty also means knowing whether 16 repetitions of an angle are needed instead of perhaps only two (or vice-versa), whether geodetic targets or range poles are better for a particular job, whether taping or an EDM should be used in a particular instance (considering both accuracy and costs), and whether a computer should be instructed to execute all calculations in "double precision" in order to avoid round-off errors. Practice of the complete art of measurement (being in control) means knowing how many figures to express in a final quantity. It also means knowing where to look for mistakes and how to apply and interpret measurement standards. Included in all of this is the need to be able to explain *why* choices and judgements are made in a particular way.

Being in control in the world of measurement means many things and knowing the theory of measurement is at the very heart of the practice of true professional level surveying.

9-3 ISOLATING AND AVOIDING MISTAKES

Without accurate data, its subsequent evaluation cannot be valid. As the saying goes—''garbage in—garbage out''. The measurer must know two important things about the job: how to control errors and how to avoid mistakes. Errors have been discussed at length. Avoiding mistakes is an art that generally starts with proper training. Initial habits learned, as passed to the technician by employers or instructors, can be the cause or direct link to many mistakes. To a large extent, understanding the equipment and the task will prevent mistakes. General mental alertness avoids many mistakes. Sloppy habits cause mistakes. A routine measurement should be done the same way each time, employing the same standard field notes and methods. For example, surveyors using a 100-ft. steel tape know how to avoid taping tally mistakes by using eleven pins and how to avoid elevation mistakes by employing standard note forms. But, it all starts with learning the best, most efficient, most clear and straightforward methods. Once the best methods are discovered through experience and experimentation, they should be rigidly employed until one learns a better way. Many surveyors and their field help continue to make the same mistakes year after year, being too stubborn to change a procedure that once worked.

Procedures and methods may need to vary somewhat with the scope of a job, the personality or physical quirks of different field personnel, weather, terrain, and equipment type or condition. But, by standardizing procedures as much as possible, many mistakes will be avoided.

The best way to deal with mistakes is to avoid them. Trouble-shooting them later is very costly and takes the fulfillment out of the work. Any employer must be fully aware of techniques and the use and limitations of the equipment and convey this knowledge to subordinates. Employers must insist upon mistake-free measurements. Nearly all mistakes can be avoided.

Many textbooks on elementary surveying describe the various surveying measurements and typically list the mistakes commonly made in each procedure. These listings are intended to provide the beginner with warnings of what can go wrong so that habits can be developed which will avoid the mistakes or so that the surveyor knows what to investigate when isolating mistakes that have already occurred. These lists are a means to become aware of possible mistakes in a particular operation and give some insight in many instances, but unfortunately they are by nature, incomplete. Developing rigid procedures designed to avoid mistakes and fully comprehending measurement as an art and science are two approaches more important than memorizing lists of typical mistakes. A third important ingredient for dealing with mistakes is a logical and clear mind. A systematic and thorough approach will find a mistake quicker than guesswork. Generally, when mistakes occur, they can be traced to 1) reading scales, 2) recording observations, 3) misusing instruments (including failure to adjust, check, calibrate), 4) measuring to or setting-up over wrong points, 5) misunderstanding processes, equipment, etc., 6) transferring data from field to calculator or computation forms, 7) manipulating numbers during computation, 8) misusing equations, 9) making programming mistakes, 10) omitting or failing to check measurements, all of which reduces to carelessness, inattention, and improper training.

It was stated earlier that it is far better to avoid mistakes than to try to locate them later. But, once made, they must be located. The magic word is to *isolate* them. Many surveyors, particularly beginners, will remeasure an entire project when a little common sense and/or mathematical analysis will help to give clues to possible location. For example, if a traverse had a large error in the eastings, but nearly perfect closure in the northings, a blunder in east-west distances might be suspected. One might investigate these first. A level circuit need not be completely remeasured if the sum of backsights minus foresights of a re-run of half the ciruicit checks with one of the previous halves. There are many other examples of this type.

One major key to quick and efficient troubleshooting of mistakes is understanding the nature of measurement and how mistakes and errors occur. In order to perform in situations involving mistakes and/or large errors, the individual must have background in measurement theory and have experience in applying it. The technician who stops understanding measurement after learning how to read the scales and keep the notes hasn't developed properly and won't be able to get out of the rough spots. Performance in tight situations is a mark of a professional in any walk of life.

Some steps in isolating mistakes are to study the notes, inquire of the personnel as to how they performed certain tasks, think back to anything that may have hurried or distracted the field personnel while measuring or recording, inspect the equipment, check the computations using independent methods, and consider the mental alertness or attitudes of the personnel and possible effects of adverse weather or work conditions at the time of measurement. Mistakes will be made more frequently if personnel have other things on their mind.

The major keys to success in finding mistakes and large errors is understanding measurements, the equipment, the procedures, and people.

As mentioned, several texts contain lists of typical mistakes in specific surveying measurements. These will not be repeated here. The following is an application of an understanding of surveying measurement in an analysis of traverse measurement mistakes. Although surveying practice certainly involves more than traversing, this illustration serves to demonstrate the use of some of the theories developed in this text and may give insight into how to apply the theories to other situations. Certainly, the reader will be provided with a systematic approach to trouble-shooting mistakes in this one important surveying procedure.

Equations were derived in Chapters 6 and 8 for estimating expected angular and positional closure of a traverse. Generally, the horizontal angular closure is checked first. Equations 6.11 and 6.12 are used for this. If the closure is higher than expected, a mistake is suspected. But, where to look without remeasuring all of the angles is the question. To find the location of a mistake in one angle, the procedure is to first compute a positional closure using the measured distances and angles, then use the orientation of this closing line to further analyze the location and size of the mistake. A single angle mistake will cause the closing line to be perpendicular to a line passing through the closing point and the point where the mistake occurred. The station containing the mistake is revealed by constructing this perpendicular. It should point to the station in question. The definiteness of this pointing will depend on how large the mistake was in proportion to other accumulated errors in the traverse. If there is some question of which station to investigate, the procedure can be repeated with the positional closure computed in the opposite direction around the traverse. The crossing of the two perpendiculars to the two closing lines thus computed should occur at or very near the station in question. An alternate way to help isolate the station most likely containing the mistake is to divide the positional closure by the angular closure in radians. The result is the distance from the closing point to the station in question. It can thus be identified by scaling. After isolating the probable location of the mistake the angle should be remeasured and the angular closure checked again. If the closure is still not within the tolerances, the analysis should be repeated to detect the next smaller mistake or error, and so on until the error of closure meets the specified or computed expected error of closure. If the station suspected contained no significant error or mistake, problems in more than one location should be suspected. It must be remembered that simply remeasuring the one measurement does not check for all possible mistakes. Correct identification of points, proper scale-reading procedures, correct use of equipment and accurate reduction of data must be checked. Figure 9.1 illustrates finding traverse angle mistakes.

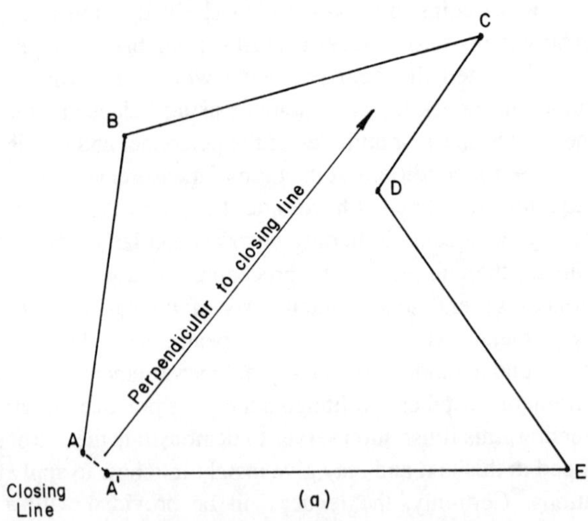

Figure 9.1 (a) Analysis of Traverse Angle Mistakes

Continuing the traverse mistake analysis, suppose that when the angular closure was checked, it was within specifications or expected closing error, but when positional closure was computed, it failed to meet specifications or expected closure according to Equ. 8.10. A mistake would then be suspected in a distance. Such a mistake is usually easily isolated. It will have occurred in a traverse course generally parallel with the direction of the closing line. Its magnitude and sign should also be revealed, being approximately the size of the closing error. The definiteness in this parallelism with the closing line, as well as the magnitude of the

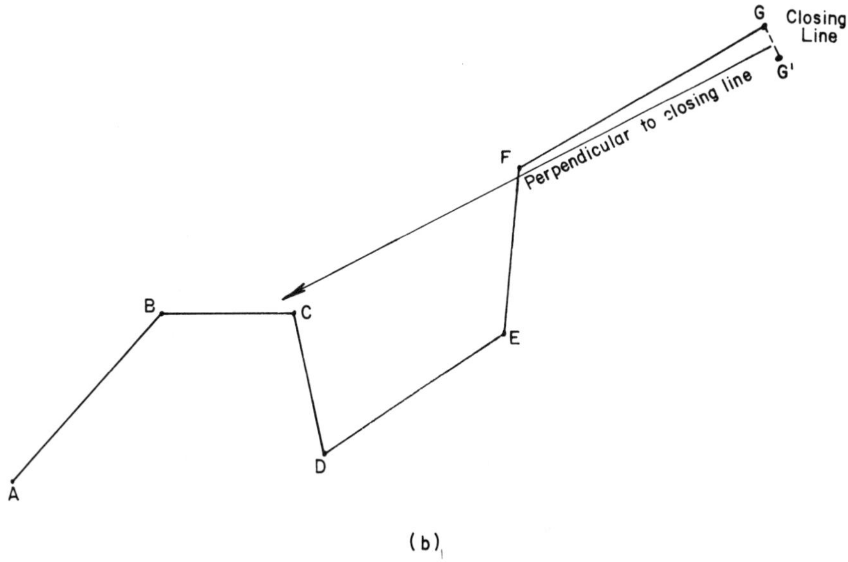

Figure 9.1 (b) Analysis of Traverse Angle Mistakes

mistake, depends on how large the mistake was in proportion to other accumulated errors in the traverse.

Often, in the above analyses, there is a possibility that a small mistake will not reveal itself readily. In such cases, the surveyor must revert to logic, to recollections of field conditions, to discussion with field personnel, and sometimes to checking several stations until the small mistake or poor work is found. But, analyses such as the above will usually save much time and effort if applied systematically. Needless to say, mistakes must be found and the quantities remeasured independently, never assuming that one has been found until actual remeasurement data are compared with the original data. Furthermore, the wise surveyor will take corrective measures to prevent similar mistakes from occurring in the future. Modifications of procedures or retraining is sometimes appropriate. A true professional will learn from mistakes.

9-4 DERIVATION OF SPECIFICATIONS

Nearly all of the discussion in previous chapters was related to the evaluation of uncertainty in measurements. The result in each example was the expected random error of the final measurement, possibly accompanied by a level of certainty of the error. The theory can be applied in reverse, starting with the desired error with its level of certainty, the solution of the problem then being a derivation of the number of repetitions, type of equipment, or care that must be exercised in each step of the measuring process. The procedure is a true design and is one of the areas of professional practice where a surveyor uses ingenuity or engineering to arrive at a process or procedure which can be applied by relatively skilled but non-professional personnel, and which has a high level of certainty (comparable to "factor of safety" in any design) of success. The design requires application of the theory of measurement, as well as judgment. It is an iterative process, starting with assumptions, then examining results and making modifications according to these results and added judgments and assumptions.

To illustrate the importance of such design, suppose a client needs a construction control point fixed in position with respect to a distant point and specifies the tolerance required. Many surveyors would arbitrarily select equipment and methods and then merely hope that the tolerance was achieved, blindly relying on least count and other elementary precision concepts. Furthermore, such surveyors would certify to the accuracy, rationalizing that they had done their best or what any prudent surveyor would have done in like circumstances, and further rationalize that this sort of project was the very reason they carry liability insurance. If tested in a court of law, there would be no accepted theory for a defense, other than the "average, prudent surveyor" actions. A more professional approach, and one that could be defended by expert statiticians and surveying scientists the world over would be a thorough design of measurement specifications based on error theory and sound assumptions. The following examples will illustrate the basic approach to design of specifications.

Example 9-1 Given: The direction theodolite of Ex. 6-1 and 6-2 having $\sigma_r = \pm 0.65$ seconds and $\sigma_{rp} = \pm 1.2$ seconds. Considering only reading and pointing errors, how many repetitions of an angle must be made to have ± 1 second error in the mean angle with 68.3% certainty?

Solution: From Ex. 6-7,

$$\sigma_{\alpha_{rp}} = \pm 1.7 \text{ seconds}$$

for one repetition of the angle.

$$\sigma_{\alpha_m} = \frac{\sigma_{\alpha_{rp}}}{\sqrt{n}}$$

where

$$\sigma_{\alpha_m} = \pm 1 \text{ second as given,}$$

from which

$$n = 2.89.$$

But, since n must be an exact number and furthermore since an even number of observations must be made with the telescope direct as inverted,

$$n = 4 \text{ repetitions (2 sets)}.$$

Example 9-2 Given: The same situation of Ex. 9-1 except that a 90% certainty is desired in the ± 1 second precision. How many repetitions of the angle are needed?

Solution:

$$E_{90_1} = 1.6445\,\sigma_{\alpha_{rp}} = \pm\,2.8 \text{ seconds}$$

$$E_{90_m} = \frac{E_{90_1}}{\sqrt{n}}$$

from which $n = 7.82$ or 8 repetitions (4 sets).

It is noted that the above answer can be calculated by multiplying 1.6445^2 by the answer to Ex. 9-1. It is seen how a higher level of certainty for the same precision requires more repetitions. It should be observed how repetitions would also increase with a lower precision specified (i.e. ± 0.5 seconds in this example). When repetition requirements become impractically high, an instrument of higher tested precision must be selected, thus lowering the numerator on the right side of the equation, resulting in lower repetitions needed.

When solving a problem such as the above for a repetition instrument, Equ. 6.3 and 6.6 are employed. The solution for *n* requires solving a quadratic equation.

Specifications would need to be prepared for angle work with a consideration for errors in target, instrument, and bubble centering, and other errors in addition to reading and pointing. These several error sources were considered in Ex. 6-8 where the uncertainty in 3 sets of an angle was determined. In that example, the uncertainties due to target and instrument centering were significant, not only because of their size, but also because they are systematic in effect. No matter how many repetitions had been made, the uncertainty in the angle would not have been much less. In fact, an infinite number would have still yielded an uncertainty of ± 2.05 seconds in this example. It is noted that merely one set of angles would have yielded ±2.77 seconds uncertainty. Therefore, the error in any number of repetitions in that field situation would have yielded about the same uncertainty because it is constrained by the systematic effect of the instrument and target centering errors. In that particular situation, the most effective way to reduce the error would be to extend the lines or use more care in centering both the instrument and targets. Often in the case of geodetic surveying, specifications give minimum sight distances for the reasons revealed in Example 6-8.

Thus, examples such as 9-1 and 9-2 are purely academic as there is much more than repetitions to consider in specifying field procedures. The next example illustrates a more realistic situation.

Example 9-3 Given: The precision specified in Ex. 9-2, being ± 1 second with 90% certainty, using the same instrument, with target distances of 1,000 feet and 2,000 feet. The vertical angle is +2° 14′ to the 1,000 foot target and −5° 30′ to the 2,000 foot target. The plate bubble has a sensitivity of 20 seconds. The horizontal angle is approximately 30 degrees. Write a specification for field personnel to follow which will yield a 90% error in the mean angle of ± 1 seconds.

Solution: It is noted that this situation is a combination of several things in Ex. 6-4, 6-6, 6-7, and 6-8. Some of the formulas in those examples will apply to this solution. It must be decided how close the instrument, the targets, and the bubble must be centered, as well as how many repetitions must be made. The reading and pointing errors are fixed unless a selection of instruments can also be made. Let us say that this is the only instrument available. From the previous discussion of Ex. 6-8, it is clear that the target and instrument centering must be improved. Let us specify a centering error of ± 0.05 inches for a first trial. Using previous solutions from Chapter 6, the individual uncertainties would be

$$E_{\alpha_t} = \pm\, 0.96 \text{ seconds} \quad \text{(See Example 6-4)}$$
$$E_{\alpha_c} = \pm\, 0.38 \text{ seconds} \quad \text{(See Example 6-6)}$$

The square root of the sum of the squares of these two errors is greater than one second. Therefore, the care in centering must be improved even more. Considering an optical plummet at each target with geodetic targets on tripods, one might be able to center to ± 0.02 inches with 90% certainty, provided the optical plummets have been checked and adjusted very carefully. With this specification,

$$E_{\alpha_t} = \pm\, 0.38 \text{ seconds,}$$
$$E_{\alpha_c} = \pm\, 0.15 \text{ seconds.}$$

The other uncertainties are reducible by added repetitions. Assuming that a bubble centering error of ±½ division is realistic, that uncertainty remains as

$$E_{\alpha_B} = \pm\, \frac{1.04}{\sqrt{n}} \quad \text{(See Example 6-7)}$$

The reading and pointing error is

$$E_{\alpha_{rp}} = \frac{\pm\, 2.79}{\sqrt{n}} \quad \text{(See Example 6-8)}$$

$$\therefore E_{\alpha_m} = \sqrt{0.38^2 + 0.15^2 + \left(\frac{1.04}{\sqrt{n}}\right)^2 + \left(\frac{2.79}{\sqrt{n}}\right)^2} = 1$$

from which

$$n = 10.6$$

The next higher even number would need to be used,

$$\therefore n = 12 \text{ repetitions (6 sets)}$$

The written field specification would be:

1. Use direction theodolite having $\sigma_{rp} = \pm 1.2$ seconds.
2. Check and adjust optical plummets for theodolite and targets. Adjustment must assure true verticality of ± 0.01 inches at 5-ft. tripod height.
3. Center theodolite and targets over stations to ± 0.02 inches or better, using geodetic targets and normal tripod heights.
4. Measure 6 sets of the horizontal angle, changing circle setting by 30° with each set and keeping plate bubble centered ± ½ division at all times. Invert telescope as per usual procedures.
5. Avoid settling ground, wind, heat waves, or horizontal refraction conditions.
6. Calculate the 90% error of the angle when completed to assure that specified precision has been met. Reject and repeat measurements as necessary. Use set averages for the computation.

This example serves to illustrate what is required in order to achieve 1-second precision from a 1-second theodolite when sight distances average much less than ¼ to ½ mile. It should be noted that the repetitions could have been lowered somewhat if more care in bubble or tripod centering could be attempted. The professional judgment of the surveyor is required and it should be carefully considered whether better precision is attainable with available equipment and human limitations. It is foolish to write specifications which are unrealistic.

In the specifications as written, n was high enough to warrant checking the precision by a calculation for σ. If n is much smaller, other rejection criteria, such as the "maximum" error (3.29 σ or 2 E_{90}) would need to be used. This would mean any set average lying more than 2 or 3 seconds from the mean of all sets would need to be rejected and the set remeasured.

In that effects of adverse conditions are impossible to quantify, specification 5 merely suggests avoiding such conditions. The changing of the circle by $180/n$ between sets is normal practice to distribute any circle graduation errors around the circle. Naturally, the two angles in a set are measured with the telescope in both positions to compensate for systematic instrumental errors discussed in Chapter 6.

In practice, a specification would not normally be written using the specific conditions at each station as a basis. It may be appropriate to write specifications for various classifications of precision desired, as a function of average line lengths, terrain as it affects vertical angles, and angle size. The vertical angles, it is recalled, affect uncertainties due to bubble centering. All else equal, a different specification might be written for hilly terrain than for flat terrain. The number of lines in a closed traverse affects average angle size and this affects uncertainties due to instrument centering. Therefore, a slightly different specification might apply to a four-sided figure as one of 50 sides or a traverse between two control points where all angles are near 180°. Naturally specifications will vary considerably with line length and instrument precision as has been demonstrated.

The surveyor should learn to prepare specifications for various measurement situations. The above iterative process, employing error theory and judgment, can be used for any type of measurement, be it angles, distances, elevations, or indirect measurements. Before certifying to any accuracy or positional tolerance, specifications must have been prepared and followed, or else the certification is merely an exercise to fulfill some legal requirement or traditional practice.

9-5 ADOPTION OF MEASUREMENT STANDARDS

As was seen in the previous section, derivation of measurement specifications starts with a decision of the accuracy required (or desired) in the result. This result could be called the "standard" to be achieved. It is often also called the "tolerance". Another familiar term is "order of accuracy". All such terms refer to overall quality of a survey and are expressive of this quality. It is important to understand that quality is not achieved without execution of all measurements according to specifications.

Computation of a mathematical closure, such as for angles or positions in traverses serves to check for mistakes or large errors caused by failure to apply specifications properly. But closure checks alone do not assure accuracy. "Closure" is not synonymous with "order of accuracy", unless it is *expected* closure, derived from design based on error theory as explained in previous chapters in this book. Arbitrary selection and use of equipment and methods could yield *actual* closures comparable to those achieved through use of well designed specifications. This could occur merely by chance, the various random errors compensating in a manner not in accord with usual theory. Faulty procedures can also sometimes fit into a pattern and be undetectible by closure computations. Suppose, for example, that field personnel failed to make slope corrections and that the east-west lines of the survey were sloping the same degree. Such a mistake in procedures would not be detected by merely checking traverse closure. An even better and somewhat classical example of how closure checks do not check accuracy is the case of a long or short tape or error in EDM calibration. Constant systematic errors are not detected when surveys close on the same point. The importance of controlled measuring procedures should be obvious. Without exercising control in every step of the measuring process according to a sound design, there is a good chance that an actual closure might be achieved but that one or more intermediate points or measurements could be significantly in error.

Accuracy standards can be stated in terms of closing error if *expected closure* rather than actual closure is used. The concept of positional tolerance is frequently used in measurement standards, and, as with closing error, has meaning only if based on expected positional error as assurred by use of designed specifications. Then, the positional tolerance of one point with respect to any other is easily computed using the same procedures for expected closure. For traverses, this has all been explained as ''positional uncertainty'' in Chapter 8. For situations other than traverses, the uncertainty (or designed tolerance) can be computed using the error theory developed throughout this book.

Many state surveying societies and licensing boards have felt a need to adopt standards of practice in order to guide surveying practices and serve as a means to police the profession and protect the public from the incompetent. Parts of the standards involve procedures other than measurement and often can be developed by consensus of opinion on what constitutes ''minimum standards''. The part involving measurements is often the most difficult to prepare, due largely to lack of understanding of measurement theory and particularly the concepts of closure, order of accuracy, and positional tolerance. Confusion often exists when it is realized that concepts often presented in such standards are incomplete and could not be checked or defended.

It should be realized that a standard can be prepared for each individual measurement, be it for distances or angles, and positional accuracy is assurred in the process. Components are easier to conceptualize than positions which are a product of the components. But, if positional tolerance is to be used, it should be easily understood as positional uncertainty. It should be clearly explained in any standard adopted that such tolerance or uncertainties are not achieved, nor should any certification be made that they have been, until specifications have been designed and used to assure that the standard has been met.

Positional tolerance or positional uncertainty must always be stated with a percent certainty (such as 90%, 95%, etc.) and be *relative* to something fixed in position. Statements are incomplete if a reference such as a control point or coordinate system origin is not given. For example, it is incomplete to state that ''positions are correct ± 0.02 feet'' or ''points are accurately located ± ...''. It is theoretically impossible to achieve such implied absolute accuracy in three dimensional space having no origin and with an implication of 100% certainty. Therefore, such a statement is meaningless, creates much confusion, and is impossible to check. Chapter 8 gave procedures to derive positional uncertainty of interrelated stations. Such theories must be used in standards of practice to lend meaning to them.

9-6 MEASUREMENTS AS SURVEY EVIDENCE

As has been stated, no notation of a measured quantity is complete without expressing the range of uncertainty, since exactness is impossible. As Gordon Gracie stated:

> Two questions of importance arise when a measurement is made: (1) How large is the quantity measured? (2) How accurate is its determination, ...No statement regarding a measurement is complete

> unless both of these questions are reasonably answered... The second question..., is unfortunately ignored all too frequently, yet basically, it is just as important as the first question. (Ref. 6 pp. 423-428.)

Measurements are an important part of the survey evidence which must be analyzed by a surveyor performing a retracement survey. Just as good monumentation enhances preservation of evidence in the original survey, so does good measurement. And, as with monuments, the more fully they are described, the better they are as evidence for the retracement surveyor. And, the better the evidence, the lower the costs for future surveys. There is no better way to protect the public welfare than to leave complete evidence. If good measurement accuracy is achieved, part of its value is lost unless the estimated uncertainties are expressed in some way. Without such notations, the user of the survey has no sound basis to compare duplicate measurements during retracement or to check for faulty work in the event of an incompetency hearing. It should be reasonable to expect that a professionally licensed surveyor is in complete control of measurement, beginning with an understanding sufficient to design and execute the measurements with a control over uncertainty and ending with a complete description of each quantity measured.

Noting the size of a measured quantity is rather simple. Noting uncertainty is not so simple. It should be recalled that without an uncertainty statement, the precision implied is half the last decimal place expressed. For example, a mile, expressed to hundredths of a foot results in an implied relative precision of about one part in a million. A note such as "section line miles were measured to an accuracy of ± 0.5 feet with 90% certainty" would imply a relative error of one part in 10,000. Such a note overrides the precision to ± 0.005 implied by expression to hundredths and permits expression of the values to tenths or hundredths of a foot for the purpose of avoiding round-off errors in computations used to check closures, make land divisions, etc. It must be reiterated and emphasized that such statements of accuracy are proper only if they are, in fact, true — that is, the surveyor must have designed and used a method to yield the accuracy stated.

There are various ways to describe measurements without unnecessarily cluttering land descriptions and plats of survey. In the first place, land descriptions should *always* refer to a plat with a note such as:

> "...all as shown on the attached plat of survey, said plat with its legend and descriptive notes being considered an integral part of this written description for survey purposes, and not to be detached herefrom."

Then, the plat itself can contain the same dimensions and area as cited in the written description and also include notes such as the one previously cited for the mile measurement. Other examples of possible statements (based on actual design or sound judgment, of course) on survey plats might be:

1. All distances have a 95% error of one ten-thousandth of the value shown.

2. Distances have been measured by electronic methods, and after corrections to horizontal datum, have a 90% error of ± 0.04 feet plus

D/500,000, where D is the distance shown.

3. Directions refer to the state plane grid meridian as determined astronomically and corrected for Laplace value and mapping angle, all directions having an estimated 90% error of ± 10 seconds with respect to this grid meridian.
4. All positions are located to a tolerance of ± 0.02 feet with respect to the point of beginning of this survey, using measurement methods designed to yield 90% confidence.
5. The area shown has been determined indirectly from distance and direction measurements and is significant to four decimal places as shown.

Such statements are certifications based on professional judgment and are excellent guides for future use of the survey work. As related to standards of practice, all but the fourth one are rather straightforward and checkable. Anything based on positional error is once removed from the dimensions and therefore more difficult to check. For this reason, standards of measurement and statements of accuracy are probably better to be framed in the context of component dimensional uncertainties.

9-7 SURVEYING MEASUREMENTS IN THE FUTURE

A Historical Perspective

This text, and particularly this chapter would be incomplete without including some historical perspective of the surveyor as expert measurer, extending what was through what is into what to possibly expect as a future role. The brief history of surveying presented in Chapter 1 is incomplete regarding where measuring has been and where it is headed. The following is intended to complete the picture.

From ancient times until the middle part of this century, there was very little difference in the principles behind the design, construction, and application of surveying instruments. Their precision and a few accessories have varied, but they have historically depended largely upon human manipulation and skill to yield accurate results. For example, rods, ropes, lines, cords, chains, and tapes all work on the same principle—they must be laid end for end, aligned, leveled, and read at the last point. Leveling has been done using water in troughs or tubes, or by a plumb bob on a string. In the seventeenth century, the telescope began to replace the peep sight on leveling instruments, otherwise the use of the instruments was basically the same—observe readings on sticks or rods from an instrument mounted on a staff or tripod. There is evidence that at least one ancient angle measuring instrument had horizontal and vertical circles with clamps, sighting devices, and motion screws—much like modern theodolites. Angle measuring instruments have always employed circles, acting like protractors, with measurements being made by centering over one point and pointing through peepsights or telescopes to other points.

The main observation here is that the traditional personal, instrumental, and natural errors have been involved in all such measurements. Although early surveyors, and even some in this century, did not usually concern themselves with error analysis and design of standards and specifications, the errors did propagate much the same in the measurements, nevertheless. The errors were surely much larger in older, cruder instruments, but good closures were not as important in the past as they are becoming now; therefore, there was less concern for learning measurement theory and its application to practical problems. With today's land values and need to interrelate many adjoining land systems, there is an increasing need to learn the true nature of measurements. That is the principal gap filled by this text — to teach the principles of surveying measurement as applied to commonly used instruments which all depend to some extent upon humans to manipulate and read, and which are subject to recognizeable instrumental and natural errors caused by easily understood mathematical and physical laws.

However, as has been outlined to some extent herein, many of the instruments have become more automatic and obey physical laws not so straightforward in analysis by the equipment user. Many obey laws related to electronics, for example. Complete understanding of the instruments and errors requires a level of knowledge beyond merely understanding coefficients of expansion or equally simple geometric laws. Furthermore, the human error has been removed from consideration in many ways. For example, the user of an automatic level has no concern for bubble centering error and the EDM user need not be concerned with controlling sag or tension, or reading a fractional tape distance.The de-emphasis of human error has been accompanied by increased inherent instrument precision, so that control of all errors in some instances is at the mercy of the developer and manufacturer of the equipment rather than the user. This means that, although the equipment and methods as outlined herein are far from being obsolete, as the decades pass by, the applications will continue to change and the variables to be controlled and considered in the propagation of error evaluations will change accordingly. The important developments of the last few decades, beyond the traditional instruments and methods, will be discussed in the following paragraphs, each as regards its impact on present, as well as foreseeable future surveying measurement, and the role of the surveyor as expert measurer.

Here, Now, and Soon

Electronic distance measurement has been discussed in previous chapters. It should be obvious that this system has revolutionized much of modern practice. Indications are that most surveyors now use such systems in everyday practice and that EDM has replaced the tape to a large extent. Thus, the consideration of many errors in distance measurements is changing. It is unlikely that the tape will become completely replaced in the foreseeable future since certain short distances and the inexpensive nature of taping itself still demands that it be used for many situations. Thus, surveyors should still retain the fine art of accurate taping in their education and training, even though they may eventually use it only a fraction of the time. If skill is neglected, the possibility of blunders and uncontrolled errors would be very high whenever it is used.

Optical theodolites are not a very recent development, but are so often considered as "modern" by so many present practitioners that they need to be discussed further. These instruments were emphasized in this text and are considered to be the primary angle measuring instrument today. Indications are that the majority of the surveyors throughout the world use and prefer optical instruments. Recent improvements in some instruments employ what are commonly called "digital readout" systems, eliminating one source of the reading error in aligning micrometer scales. However, the accuracy or resolution is not significantly better than what the human can accomplish in aligning marks and reading scales. With such instruments, other errors such as pointing, eccentricity, and the various instrumental errors are still present. On the horizon, however, are systems which, at touch of a button will display horizontal and vertical angles (directions), which are also recorded onto tapes. Even with these systems, other error sources are still present. Reading errors are also still present but are instrumental rather than personal. It is conceivable that reading errors could be lower than on conventional theodolites. However, the danger in trusting a highly precise reading system is that, just as now, the other error sources may have much greater influence on total angular precision, and the user could be deluded into thinking the reading precision indicates the system precision.

Total station is a term commonly used for a system now in the production stage which generally employs nothing more than the usual EDM and theodolite, but all mounted on the same tripod simultaneously. At least one such model has digital displays of directions as well as distances. Such units generally display horizontal distance, a computer unit having corrected the slope measurement for observed vertical angle. Thus, some of the errors, time required, and possibilites for computation mistakes are eliminated. Some such units compute, display, and record elevation differences as well. The principal advantage of such instruments are speed in gathering data. Most of the error sources in such distance and angle measurements are still present. At the writing of this book, many of the total station packages were beyond the price range of most surveyors, especially the more precise direct-readout units, and therefore have not been widely employed.

Leveling hasn't changed much. Obviously, the automatic level has eliminated the level bubble and the personal error in leveling and replaced it with an almost negligible instrumental error. On the horizon is the production of a level instrument employing a free mercury surface as a mirror for leveling the line of sight. The possible advantage over the pendulum type of automatic level is that mercury levels itself very quickly and it is stable. There are other advantages over bubble type tilting levels related to effects of natural errors in fluid expansion of bubbles and shorter time in the fluid coming to rest. There is no apparent difference between such an instrument and those already discussed as regards error propagation. The line of sight uncertainty would need to be evaluated for each particular instrument.

The only instrument on the horizon that could revolutionize leveling is in the category of inertial surveying systems, which will be discussed subsequently.

Doppler satellite techniques are a space age technology employing launched satellites to determine precise geodetic position by signals between the orbiting satellites and the earth. At present, this technology is not feasible for the practicing surveyor, and it may not soon be advantageous for engineering and land surveyors. It is primarily a geodetic control system, but also has military applications. The National Aeronautics and Space Administration, the Defense Mapping Agency, the National Geodetic Survey, and private organizations have been involved in research and development of these systems. Present systems employ "geoceivers" which receive radio waves transmitted from the orbiting satellites. Changes in frequency at widely spaced ground stations yield data for determination of ground positions. Positional accuracy up to one meter has been claimed. The techniques are still too expensive, inaccurate, and cumbersome for most surveyors to employ. At the writing of this text, the systems, although being applied, are still in the development stage.

Inertial systems incorporate gyroscopes, accelerometers, and computers to determine positions on the earth.This is a system which comes closest to the surveyor's fantasy of a little "black box" for instant determination of a position in the three elements of space. The system operates by a process called "gyrocompassing". Two gyroscopes orient themselves by sensing the earth's rotation, one pointing north-south and the other east-west, becoming aligned with respect to the direction of gravity. Accelerometers measure relative movements as the instrument is moved between ground stations. By simply orienting or calibrating on a point of known position, absolute position of other stations are determined. Absolute positions have been determined to accuracies less than 1 meter. Again, these systems have been researched and developed for military, as well as civilian purposes. Systems have been tested and used by the Defense Mapping Agency, the U.S. Bureau of Land Management, and the Geodetic Survey of Canada.

Some surveyors have stated that this system has much potential for completely revolutionizing current surveying practices and could make the traverse itself obsolete in a few decades. Other surveying scientists state that these systems may not have an impact very soon on surveys for engineering and property boundaries because of the higher accuracy needed for such surveys. However, development of inertial systems will eventually affect equipment and methods for all surveys.

Photogrammetry, in its simplest meaning, is the art and science of making measurements using photography. Thus, if photographs are of the earth's surface, it can be seen that photogrammetry is another surveyor's tool for land measurement. This science has been with us much longer than many realize, it being as old as photography itself, dating back to the mid-nineteenth century. Important research was done in photogrammetry during the early part of the twentieth century and by the middle of this century, it had mostly replaced conventional ground methods for production of topographic maps, particularly small-scale maps of large areas. Development of high-speed electronic computers has caused another surge in the application and importance of photogrammetry as a measuring and mapping tool. Its potential and present-day applications for determining relative positions in the three elements of space is often unrecognized by present practitioners. Accuracies to a small fraction of a meter are possible using certain analytical techniques. Such accuracies are sufficient for many surveying purposes.

Many of the theories introduced herein are applicable to the analysis of photogrammetric measurements, the science being largely one of geometry. Photogrammetry is a method of indirect measurement, since positions are determined from stereoscopic models, distances and directions being determined indirectly from coordinate read-out. Although this science will not fully replace most direct field surveying measurements, its incorporation as a valid and reliable measuring tool by ordinary surveyors should increase steadily as education of surveyors improves.

Computers and calculators deserve mention as their use is related to errors and their analysis and control. Concern over propagation of round-off errors is reduced considerably with widespread use of electronic computers and calculators since trigonometric functions and typical conversions and constants do not need to be read from handbooks and printed tables. Such systems are here and are accepted. The danger of such high speed and precise calculation systems is that false security can be present, the user sometimes not realizing that original measurement data fed into computers or onto the calculator keyboards must be accurate. It must continually be emphasized that computation precision cannot improve the precision of the data; it can only maintain the degree of precision entered into the computer system. Likewise, it cannot correct mistakes and unless people who can interpret the results interact and check results, the mistakes will remain.

Remote sensing surveying is often considered a new branch of photogrammetry. In that sense, it is another measuring tool for surveyors, with one distinct difference— it is primarily ''environmental surveying'', rather than being involved with only the three elements of space and their measurement. Remote sensing utilizes methods such as color and color infrared photography, thermal scanners, and radar to collect data on the earth's natural and cultural resources and to detect and monitor the impact of man's use of the land. Data are useful in city and regional planning, conservation and use of natural resouces, and as a general record of the land. Data are typically collected from aircraft or satellites. Measurement, in the conventional sense, is inherently involved in the process, since the data must be mapped and positioned. Locations must be determined with respect to known ground systems. Therefore, surveying measurements and their analysis are involved in order to accurately plot the data spatially. Measurements are also involved in the interpretation of the data itself, photogrammetry being involved, as well as measurements with densitometers and other instruments.

Remote sensing data are used by many professionals outside of the surveying profession. Proper data collection and analysis can only be done by those familiar with survey measurement theory. Although this relatively new branch of surveying will not change measurement science itself, it merely adds another dimension of concern for the measurement scientist.

Effects on Measurement Analysis

The theories and occurrence of random and systematic errors remain unchanged, despite recent and potential future changes in measurement technology. As applications of instruments change, applications of the theory will change with it. Some measurement error controls will become less important and some may become more important. Surveyors will still need to know the error sources and how to compensate or minimize their effects. They will still need to deal with computations to divide land and make other indirect measurements. They will still be concerned with the quality of computer data input and analysis of output.

Angles, distances, and conversions will always be of concern, as will delineation of real estate, construction controls, and geodetic control on the ground. Although inertial systems and possibly other automated instruments may change practice, the analysis of the measurements and their accuracy will still remain largely in the hands of educated surveyors who understand both the art and the science of measurement.

Problems - Chapter 9

9.1 What does "being in control" of surveying measurement mean in a professional sense?

9.2 Discuss how a surveyor can "be in control" regarding measurement mistakes.

9.3 The following traverse data contains a mistake:

Line	Azimuth	Distance, ft.
AB	31° 30′ 30″	471.85
BC	100° 42′ 15″	552.71
CD	184° 10′ 35″	268.93
DE	179° 58′ 40″	285.09
EF	269° 58′ 20″	760.48
FA	357° 55′ 40″	254.08

The traverse was measured using methods designed to yield 1:10,000 relative error of closure. Find the probable location of the mistake, its magnitude, and in which component it occurred.

9.4 The following traverse ABCDEA contains an angle misclosure as the field procedures should have yielded an angular error of closure of not more than 30 seconds.

Station	Measured Angle	Line	Distance, ft.
A	130° 16′ 48″	AB	2529.60
B	62° 41′ 12″	BC	5210.81
C	38° 42′ 56″	CD	4073.75
D	116° 02′ 07″	DE	482.12
E	192° 12′ 02″	EA	451.16

Find the probable magnitude of the mistake and its most likely location.

9.5 Derive a set of specifications for a differential leveling loop of 2 Kilometers length which will yield a closure of at least 5 mm$\sqrt{K_m}$ where K_m is the distance in Kilometers, and assuming an automatic level with compensator sensitivity of ± 0.5 seconds is available.

9.6 Using the configuration of the traverse of Problems 8.4 - 8.7 but neglecting the error estimates given, how precisely must the distances and directions be determined in order to yield an expected closure of 1:25,000, maintaining consistency in distance and directional accuracy?

9.7 In establishing a taped baseline, it is decided that the maximum tolerable uncertainty per 100-ft. tape length, due to sag of the tape is ± 0.002 ft. Given that the tape weight is 1.6 lbs., and that tension can be determined ± 2 lbs., what minimum tension must be applied to the tape?

9.8 Suppose that the same maximum tolerable uncertainty per tape length of Problem 9.7 also applies to uncertainty due to temperature. How closely must temperature be determined?

9.9 Derive a taping specification designed to yield a relative error in distances of 1:10,000 over distances of 1 to 2 miles using a 100-ft. steel tape over flat to gently rolling terrain.

9.10 Assume that differential levels could yield an expected error in an elevation difference between two points of ± 2.0 cm., the points being 3.5 Kilometers apart and Δ Elev = 125.2 meters. What precision would be required in measurement of a vertical angle if the elevation difference was to be determined with comparable precision by trigonometric leveling, and assuming the error in an EDM distance for this purpose is negligible?

Additional Reading References - Chapter 9

1. Andregg, Charles S. ''Instruments and Methods for Surveying and Mapping'', *Surveying and Mapping,* June, 1977.

2. Andregg, Charles S. ''Where Is Technology Taking Us?'' *Surveying and Mapping,* December 1977.

3. Barry, Austin B. *Errors in Practical Measurement in Science, Engineering, and Technology.* New York: John Wiley & Sons, Inc., 1978, Chapter 9, Appendix C.

4. Brown, Curtis M. and Eldridge, Winfield H. *Evidence and Procedures for Boundary Location.* New York: John Wiley and Sons, Inc., 1962, Chapter 16.

5. Buckner, R.B. ''Preserving and Perpetuating Land Survey Evidence'' *Surveying and Mapping,* December, 1972.

6. Gracie, Gordon ''Reliability of Field Measurements in Property Surveys'', *Surveying and Mapping,* Sept. 1963.

7. Hodgman, Francis *Land Surveying.* Columbus, Ohio: CARBEN Surveying Reprints, 1976 (1913 original), Chapter 12.

8. Kivioja, L.A. ''New Mercury Leveling Instruments'' *Surveying and Mapping.* March, 1979.

9. Mikhail, Edward M. and Gracie, Gordon. *Analysis and Adjustment of Survey Measurements.* New York: Van Nostrand Reinhold Company, 1980, Chapter 7.

10. Mueller, Ivan I. and Ramsayer, Karl H. *Introduction to Surveying.* New York: Frederick Ungar Publishing Company, 1979, Chapter 1, Sections 4.42 - 4.44, 6.4.

11. Wolf, Paul R., ''Surveying: Current Status and Future Challenges'', *Surveying and Mapping,* June, 1976.

CHAPTER 10

INTRODUCTION TO ADJUSTMENTS OF MEASUREMENT DATA

10-1 BASIC PURPOSE AND THEORY OF ADJUSTMENTS

Whenever more measurements are made than are necessary to determine a quantity, *adjustments* are appropriate. Two measurements of an angle for example, result in one measurement more than is absolutely necessary to establish a value. Computation of the "best value" (arithmetic mean) is a simple adjustment which encompasses most assumptions required for complex measurements. Extra measurements are called *redundant measurements*. Higher redundancy increases the complexity of the adjustment since each additional measurement imposes new *conditions* that must be satisfied in order to resolve discrepancies caused by propagation of random errors. For example, in triangulation, a chain of triangles is easier to adjust than a chain of quadrilaterals because the additional angular measurements made in the quadrilateral add more geometric conditions. But, the additional conditions add precision to the adjusted values. Professional surveyors always have a judgment to make regarding just how many redundant measurements are actually necessary in order to sufficiently check for mistakes as well as have adjusted values with sufficient precision.

The purpose of any *adjustment* (also, sometimes called "balancing") is to distribute closing errors throughout the survey so that lengths, directions, angles, heights, or positions have values as close to true as analysis of the random errors will permit. Adjustments relate to random error theory. The first basic assumption is that all systematic errors and mistakes have been detected and removed — only random errors should remain before adjusting. A second basic assumption, related to the first, is that the closure required by specifications must have been achieved before adjustments are made. Unfortunately, adjustment techniques can be applied mathematically whether closure requirements have been met or not. It is foolhardy to apply elaborate adjustment methods before closing errors are within tolerances required. A third basic assumption is that the field methods used were those on which the adjustment method is based. Sometimes, for example, an adjustment program is prepared based on the relative precision of distances and angles as achieved by a certain instrument combination. If the field work does not follow the planned techniques, the adjustment method may be invalid and result in less than the best possible values for the adjusted quantities.

Adjustments are generally made in proportion to some quantity, based on the random error theory that there is an equal likelihood of the errors occuring in one place as another. For example, if three angles in a triangle failed to close by three minutes, it would be logical to add a correction of one minute to each angle to effect the closure. Adjustments in differential levels are made in proportion to the number of readings or length of the circuit, along with the assumptions that the likelihood of errors occurring in one reading was the same as in any other. The exceptions to adjusting or distributing errors evenly or proportionally are when (1) varying precision was employed in different parts of the survey, in which case, *weighted adjustments* are applied, or (2) certain field conditions create situations where errors are distributed unevenly, according to the surveyor's judgment.

Each adjusted value is computed as follows:

$$\text{Adjusted Value} = \text{Unadjusted Value} + \text{Correction} \qquad (10\text{-}1)$$

An important principle that should be applied after performing any adjustment is that the adjusted values should be checked to make sure they result in an adjusted closing error of zero. In the case of the three angles of a triangle, they should add to 180° after adjustment, for example.

10-2 METHODS OF MAKING ADJUSTMENTS

The most precise method of adjusting measurements is through the application of *least squares*. This is a method based on the theory of probability whereby the sum of the squares of all the corrections (or errors or residuals) applied to the observed data is made a minimum. The minimizing of this sum requires first defining the residuals in terms of known values, then adding the squares of the terms thus formed, then differentiating this sum and setting the derivatives equal to zero. This process will be clarified through subsequent examples. Each individual problem is different and must be analyzed in a manner which satisfies all conditions with the basic requirement that the sum of the squares of the adjustments is the least it can be. In this sense, the positions or quantities have been adjusted, collectively, as closely as possible to the true values, and their total change is minimized.

Although least squares is the best method in terms of precision and "best fit", the precision obtained by applying it is sometimes beyond the significant digits warranted by the survey data, particularly if closing errors are relatively small. The added work of applying a least squares analysis may not be justified in such situations if a simpler method is available which would yield essentially the same values. Sometimes, closing errors might even be ignored completely if they are sufficiently small, or might be arbitrarily placed in selected measurements. For example, if an angular closing error of one second was observed in a traverse using measurements of ± 1 second precision, this is so small that no adjustment is really necessary—the one second might be placed in one angle only since the significant figures in the measured values do not warrant breaking it down into tenths. Such adjustments are sometimes called *arbitrary*.

The *field method* adjustment is similar to the arbitrary method except that it has connotations of having some judgment applied as to the best location to place the errors. Using the field method approach, corrections are made as based on observations of conditions in the field which may cause a greater probability of errors having occurred in one location than another, according to the surveyor's judgment.

10-3 SIMPLE ADJUSTMENTS

The closing error of a closed figure is easily determined by geometry. For an open figure, when closing on a distant point of known azimuth on the same grid system as the starting azimuth, the closing error follows the usual formula for error:

$$\text{Closing error} = \text{observed azimuth} - \text{known azimuth}.$$

Once the closing error is found, either from the above formula for an open figure or (n-2) 180° for a closed loop, this error must be distributed among all angle points in order to have a unique value for each azimuth. If the theoretical "best" value is to result, the collective differences between observed and adjusted values should be as small as possible, taking into account any varying conditions among angle points. Random error theory dictates that there is an equal chance of errors occurring at one point as at any other point, if the field conditions and methods are consistent throughout the survey. If it is assumed that conditions are equal, then it is valid to make equal corrections. Thus, the closing error is merely divided by the number of angle points, signs changed (C = −E), and the equal corrections added to the observed angles.

Example 10-1. Given: The interior angles of a closed triangle have been measured with readings as shown. Assume that all angles were measured with equal care and precision.

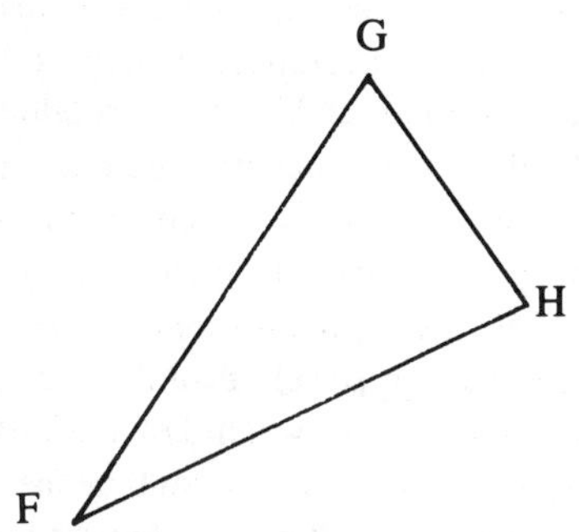

$$F = 40°\ 12'\ 13''$$
$$G = 52°\ 46'\ 02''$$
$$H = 87°\ 01'\ 41''$$

What is the adjusted angle at each station?

Solution: The sum of the angles is 179° 59′ 56″, resulting in a closing error of −4″. Each correction (adjustment for closing error), therefore, would be

$$C = \frac{+4}{3} = 1.33''$$

But, it would be indicating false precision to show adjusted values to precision of hundredths of a second. Even tenths of a second may be inappropriate. The nearest second is probably proper, but for the sake of avoiding round-off errors in subsequent calculations, let us round corrections to tenths, necessitating arbitrarily making one correction 1.4 seconds, the other two, 1.3 seconds. Actually, the configuration of the traverse leads to the possiblity of more error in station H because the sum of the sight distances there is least, and it is known that more error is possible when sight distances are shorter. Therefore, station H is given the 1.4 seconds adjustment. Thus, the final corrected values result from an equal-adjustment assumption, as well as a field condition influence, not being entirely arbitrary. The final values are

$$\begin{aligned} F &= 40^\circ\ 12'\ 14.3'' \\ G &= 52^\circ\ 46'\ 03.3'' \\ H &= 87^\circ\ 01'\ 42.4'' \\ \hline \Sigma &= 179^\circ\ 59'\ 60.0''\ \text{check} \end{aligned}$$

Many other situations of simple adjustments arise in everyday surveying practice. Measurements where redundancies exist and where comparisons are made with previously established accepted values or with geometrically true values, all require adjustments. Most of these common, simple methods are adequately covered in many elementary surveying texts and will not be repeated here. What should be understood about most such methods is that they involve certain assumptions about the data and configuration of the figure and seldom yield the least possible change from observed to adjusted values. For example, the so-called "compass rule" (Bowditch method) of traverse adjustment is valid only when distances and directions have been measured with comparable precision. It cannot account for varying precision or field conditions and ignores configuration of the traverse as a consideration. It yields disproportionately large changes in directions after adjustment for east-west and north-south lines. To overcome this, some surveyors prefer the so-called "transit rule" which doesn't change the directions of cardinal lines after adjustment. However, this system is valid only where directions are assumed to be measured more precisely than distances and has the advantage of keeping directions unchanged only when they are in fact, east-west and north-south. Such methods, therefore, rarely result in the best overall adjustment. However, when closures are small, they are often sufficient. For higher precision and in order to consider weights, least squares methods are better.

10-4 APPLICATION OF WEIGHTS

Sometimes adjustments should be made in a way that allows for varying precision in measurements or varying probability of occurrence of errors in different locations of the survey. This introduces the concept of *weight,* which is relative reliability among measured quantities. As applied to adjustments, quantities with higher weight would receive less adjustment. Weights can be assigned according to judgment, number of repetitions, precision indexes (such as standard errors), or other methods. As based on number of repetitions, weights are assigned in proportion to the numbers, the assumption being that the mean of the set with higher repetitions is more precise. The following example illustrates this method.

Example 10-2 Given: The same triangle and results of Example 10-1 except that angle F was measured 8 times, G was measured 4 times, and H was measured 2 times. What is the adjusted angle at each station, considering weights?

Solution: Weights are as follows:

$$W_F = 4\,, \qquad W_G = 2\,, \qquad W_H = 1$$

It is noted that these could have been 8, 4 and 2, but it is better to reduce the lowest one to unity, keeping the ratios the same, for the sake of computing with smaller numbers. What these weights indicate is that the correction to F should only be ¼ that of H and ½ that of G, or that H's correction is twice that of G, etc. Thus, adjustments are made in inverse proportion to the weights. Since the corrections must add to minus the closing error, the ratios of adjustment must add to unity. The following illustrates these points.

$$C_F = \frac{0.25}{1.75} \times 4'' = 0.57'' \text{ or } 0.6''$$

$$C_G = \frac{0.5}{1.75} \times 4'' = 1.14'' \text{ or } 1.1''$$

$$C_H = \frac{1}{1.75} \times 4'' = 2.29'' \text{ or } 2.3''$$

$$\Sigma = 1 \qquad\qquad \Sigma = 4.0''$$

In the above,

$$\frac{1}{W_F} + \frac{1}{W_G} + \frac{1}{W_H} = 1.75$$

and C_F, C_G, C_H are corrections at each stations. Adding the corrections to the observed values yields

$$\begin{aligned} F &= 40^\circ\ 12'\ 13.6'' \\ G &= 52^\circ\ 46'\ 03.1'' \\ H &= 87^\circ\ 01'\ 43.3'' \\ \hline \Sigma &\quad 179^\circ\ 59'\ 60.0'' \text{ check} \end{aligned}$$

The difference in these answers and those of Ex. 10-1 should be noted.

Weights can also be assigned according to a precision index. Since it is known that

$$\sigma_m = \frac{\sigma_s}{\sqrt{n}},$$

there is a relationship between this method of applying weights and the method of repetitions. That is

$$n = \left(\frac{\sigma_s}{\sigma_m}\right)^2$$

If σ_m or any other error in the mean at the points is known, rather than n, the weights are proportional to the inverse of the squares of the precision indexes. Using Ex. 10-2 for illustration, it is assumed that the method (and thus σ_s) at each point was the same. With the variation in n, however, σ_m would vary. The ratios $\sigma_{m_F}:\sigma_{m_G}:\sigma_{m_H}$ would be

$$\frac{1}{\sqrt{8}} : \frac{1}{\sqrt{4}} : \frac{1}{\sqrt{2}}$$

The ratios of the squares of the inverses of each σ_m, or

$$\left(\frac{1}{\sigma_{m_F}}\right)^2 : \left(\frac{1}{\sigma_{m_G}}\right)^2 : \left(\frac{1}{\sigma_{m_H}}\right)^2$$

are 8:4:2 which is the same weight ratio used in Ex. 10-2. Thus, if the precision indexes are known instead of repetitions, the weights are made according to the inverse of the squares of these indexes. But, actually, the adjustments are made according to the ratio of squares of the precision indexes, since the ratios of adjustment are, in turn, proportional to inverses of weights. Thus it is seen that the ratios $\sigma_{m_F}{}^2 : \sigma_{m_G}{}^2 : \sigma_{m_H}{}^2$ yield the same ratios of adjustment as in Ex. 10-2.

$$\left(\frac{1}{\sqrt{8}}\right)^2 : \left(\frac{1}{\sqrt{4}}\right)^2 : \left(\frac{1}{\sqrt{2}}\right)^2 = 0.125 : 0.25 : 0.5$$

is the same as 0.25: 0.5: 1 used in Ex. 10-2. Therefore, when using precision indexes for adjusting, weights do not actually need to be computed.

Another common application of weights is in computing a weighted mean. This is a method of arriving at the "best value" of a quantity being averaged from measurements of varying precision. The formula is simply

$$\overline{X}_w = \frac{\Sigma\, w_i\, x_i}{\Sigma\, w_i} \qquad (10\text{-}2)$$

where

$$\overline{X}_w = \text{weighted mean,}$$

$$w_i = \text{weight assigned to measurement } x_i \text{ .}$$

10-5 LEAST SQUARES THEORY

Without delving into random error theory and the equation of the normal probability curve, which is necessary for a complete theoretical explanation of the least squares principle, an understanding of the concept can be gained from reasoning through the way it works. Section 10-2 gave the basic definition of least squares. If the sum of the squares of the corrections or residuals is the least possible in any adjustment situation, the best adjustment has been made. Any other combination of corrections would result in the sum of the squares being larger. What is more important is that the total adjustments, considering the survey as a whole would be larger. This is less desirable than a least squares result. The minimum possible change to all observed quantitites is most desirable. This occurs only when the sum of the squares of the "changes" is least. Thus, the name "least squares".

The way it works can be graphically illustrated by observing what happens when several points have been found which are supposed to lie on a straight line but do not because of survey errors. Suppose there are three points A, B, C whose coordinates have been found to be:

Sta	x	y
A	0	0
B	100.00	100.00
C	200.00	400.00

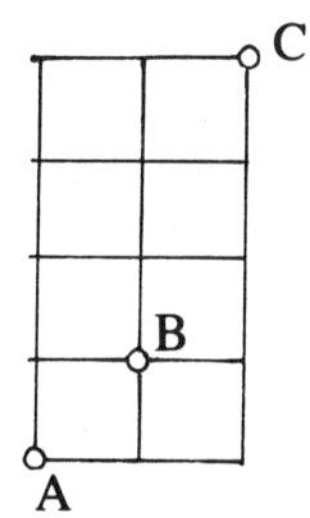

Figure 10.1 illustrates four possible solutions to fitting a straight line to these three points. The problem, of course, is that there is one more point than needed to determine a straight line. If points A and B are used, this means that point C is rejected, completely, as in Fig. 10.1 (a). In (b), point A was rejected, and in (c), point B was not considered. But, what if there is no reason to reject any point? Perhaps all are original monuments in a land survey, are undisturbed, and are equally reliable. Then, there is no basis to reject any of the three points. In other words, all should have equal consideration or *weight* and the "best fit", using all three should be computed. Even casual observation would indicate that either choice (a) or (b) results in more maximum change than solution (c), and if any single point was to be rejected, solution (c) would cause the least disruption. However, some compromise seems logical if the least collective change in the survey is to result, considering all points equal and thus being equitable as regards

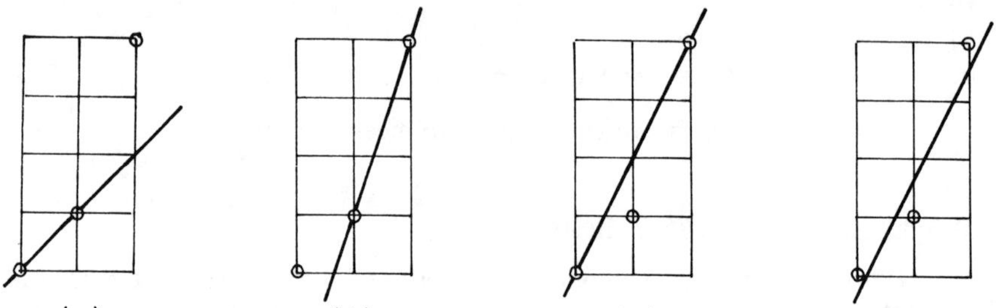

Figure 10.1 Fitting a Straight Line Using Least Squares.

land ownership or other considerations. Thus, intuitively, the average surveyor would probably select solution (d), placing the line somewhere between the points. But, exactly where? Least squares will reveal the position causing least disruption. Rough scaling of the perpendiculars from each point to the line indicates that the sum of the squares of these offsets from the line in (d) is less than the square of the perpendicular to the line in solution (c). This indicates solution (d) as sketched is approximating the correct solution. But, any minor shift or rotation from the ideal solution will result in a larger sum of the squares of these offsets. (Think of them as corrections or residuals.) To find the theoretically best solution, we can proceed as follows.

For simplicity, first divide all coordinate values by 100. The equation of a straight line is y = mx + b where m is the slope and b is the y intercept. As the line changes, the perpendicular offsets change, but so do the x and y components of these offsets. Minimizing these component offsets should also minimize the perpendicular offsets. Working with the y component of the residuals, ν ,

$$y_i + \nu_{y_i} = mx_i + b$$

is the actual equation of the line at any point. Thus,

$$\nu_{y_i} = mx_i + b - y_i$$

At the three points A, B, C,

$$\nu_{y_A} = 0 + b - 0 = b$$

$$\nu_{y_B} = m + b - 1$$

$$\nu_{y_C} = 2m + b - 4$$

and

$$\nu_{y_A}{}^2 = b^2$$

$$\nu_{y_B}^{\ 2} = m^2 + b^2 + 1 + 2mb - 2m - 2b$$

$$\nu_{y_C}^{\ 2} = 4m^2 + b^2 + 16 + 4mb - 8b - 16m$$

$$\Sigma\nu_y^{\ 2} = 3b^2 + 5m^2 + 17 + 6mb - 10b - 18m$$

To minimize $\Sigma\nu_y^{\ 2}$, its derivative with respect to each variable must be taken and set equal to zero. The variables are the slope, m, and y intercept, b.

$$\frac{\partial\ \Sigma\nu_y^{\ 2}}{\partial m} = 10\ m + 6b - 18 = 0$$

$$\frac{\partial\ \Sigma\nu_y^{\ 2}}{\partial b} = 6b + 6m - 10 = 0$$

Solving these *normal equations* simultaneously yields $m = 2$ and $b = -⅓$. Thus the best fitting straight line through the three points is

$$y = 2x - 33.33...$$

It is seen that this line is parallel to one through points A and C but is located ⅓ of the distance toward B. If desired, the perpendicular offsets to the best fitting line could now be computed. The sum of the squares of these offsets has been minimized using this least squares approach.

The primary purpose of the above illustration is to demonstrate the method of least squares and to graphically show that it works. It is to be noted that other curves, such as parabolas or circles can be fitted to data points. It should also be observed that additional points would mean additional normal equations and a more complicated simultaneous solution.

The following example demonstrates another use of least squares and further illustrates the procedure of considering weights in the process.

Example 10-3. Given: The same situation of Ex. 10-2. Solve it by the least squares method.

Solution: The condition to be satisfied is that the sum of the corrections should equal 4″, with allowance for the varying weights.

Thus,

$$\nu_F + \nu_G + \nu_H = +4''$$

where

ν_F , ν_G , ν_H are the corrections to the measured angles.

The least squares condition that must be met is

$$\Sigma\, w_i \nu_i^2 = \text{minimum},$$

which in this example is

$$w_F \nu_F^2 + w_G \nu_G^2 + w_H \nu_H^2 = \text{minimum}.$$

The weights, it is recalled are $w_F = 4$, $w_G = 2$, $w_H = 1$.

From the condition equation,

$$\nu_H = 4 - \nu_G - \nu_F\,,$$

the sums of the squares can be written

$$\Sigma w_i \nu_i^2 = w_F \nu_F^2 + w_G \nu_G^2 + w_H\,(4 - \nu_G - \nu_F)^2$$

which after expanding, rearranging and substituting the values for the weights, becomes

$$\Sigma w_i \nu_i^2 = 5\,\nu_F^2 + 3\nu_G^2 - 8\nu_F - 8\nu_G + 2\nu_F \nu_G + 16$$

$$\frac{\partial \Sigma w_i \nu_i^2}{\partial \nu_F} = 10\,\nu_F + 2\nu_G - 8 = 0$$

$$\frac{\partial \Sigma w_i i \nu_i^2}{\partial \nu_G} = 6\nu_G + 2\nu_F - 8 = 0$$

The simultaneous solution of these equations yields

$$\nu_F = +\,0.57''$$

$$\nu_G = +\,1.14''$$

The correction for H is obtained by substituting the above values for ν_F and ν_G into the initial condition equation. Thus,

$$\nu_H = +\,2.29''$$

It is seen that the results are the same as in Ex. 10-2 for the corrections, and thus the adjusted values will be the same. Had the weights in the example been equal as in Ex. 10-1, equal corrections would have resulted.

As an illustration of the field method adjustment approach, a surveyor might have looked at this situation, recognized that angle F should receive the least correction, H the most, and G somewhere between. Having only 4″ to distribute, a logical solution would have been to place 2″ correction in angle H and 1″ each in F and G. Or, perhaps another logical adjustment would have been 0.5″ correction to F, 1″ to G, and 2.5″ to H, accounting for the difference in repetitions at F and G. In either case, the corrected angles would be essentially the same as either the least squares results or the computation employing the inverse weights ratio within the significant figures of the method. Thus, it is seen that, for simple situations, elaborate calculations may not be necessary. It is emphasized, however, that many adjustment problems are not this simple and do require least squares or other computational approaches.

The advantage of the least squares method is that it gives the smallest possible changes to original observed values and affords a straightforward way to apply weights and hold certain conditions fixed. It can be applied to level circuits, fitting points to lines or curves, traverse and triangulation adjustments or any situation where there are redundancies and a means to express residuals in terms of the observations or geometric conditions. For more complex situations with a high number of redundancies, matrix algebra is a valuable tool in the solutions. Programmable calculators or electronic computers have made it possible and practical for surveyors to make least squares adjustments.

10-6 ADJUSTMENTS AND MEASUREMENT ANALYSIS

The central purpose of this text has been to lead the student to an understanding of surveying measurement. Means to estimate, predict, and control uncertainty in measurements (both direct and indirect) have been developed herein. It is felt that absorbing the concepts presented in this book is a requirement for a surveyor to achieve professional status. It is also required before the student can progress into the complexities of adjustment computations. A surveyor can analyze a project, test equipment, derive specifications, and predict and check closures employing the theory of measurement. Often, that is all that is necessary to have a complete quality job, sound on both a technical and a scientific basis. In other words, adjustments of the data are not always essential and are actually the last step in a total process. Adjustments are not part of the analysis, but are a final refinement that yields one unique value for all quantitites, considering the geometry and other conditions. Adjustments provide a convenience for future use of the data. Everything "closes" mathematically. But the values used are no more free from error than were the unadjusted values. Adjustments are folly if the project has not been properly analyzed with adequate error control and achievement of designed or expected desirable closures. It is hoped that readers will appreciate the need for proper analysis prior to adjustment. It is also hoped that this brief introduction will stimulate interest in further study of this important topic, it being an extension of the measurement analysis process and another phase toward development into an expert in the art and science of measurement.

Problems - Chapter 10

10.1 What are three basic assumptions required before adjustments of survey data can be completed?

10.2 What is the basic purpose of adjustments of survey data?

10.3 The interior angles of a closed traverse have been measured, as follows:

A	10° 12′ 35″	D	145° 36′ 35″
B	116° 18′ 45″	E	62° 05′ 55″
C	192° 46′ 10″	F	193° 00′ 30″

Assuming that the closure meets specifications, adjust the angular error of closure, list adjusted values, and show a check on the computations.

10.4 A differential level circuit has been completed with notes shown below:

Sta	BS	HI	FS	Elev	Adj. Elev.
BMA	4.61				365.02
	8.92		5.90		
	6.35		7.83		
	3.85		6.67		
	12.00		7.84		
BMA			7.54		

Compute the adjusted elevations at each station in the circuit.

10.5 A line was measured three times by different methods, yielding varying precision, values being as follows:

416.83 m ± 5 cm.
416.80 m ± 3 cm.
416.88 m ± 6 cm.

Compute the best value for the distance.

10.6 Two surveyors measured an angle using the same instruments, but one measured more times than the other. Their mean values and number of repetitions are as follows:

Surveyor A: 14° 35′ 05″ $n = 2$

Surveyor B: 14° 35′ 08″ $n = 6$

Compute the best value of the angle, considering data of both surveyors.

Additional Reading References - Chapter 10

1. Barry, Austin B. *Errors in Practical Measurement in Science, Engineering, and Technology.* New York: John Wiley & Sons, Inc. 1978, Chapter 8.

2. Mikhail, Edward M. *Observations and Least Squares.* New York: A Dun-Donnelley, 1976.

3. Mikhail, Edward M. and Gracie, Gordon. *Analysis and Adjustment of Survey Measurements.* New York: Van Nostrand Reinhold Company, 1980.

4. Moffitt, Francis H., and Bouchard, Harry. *Surveying*, Seventh Ed. New York: Harper and Row, 1981, Art 4-8 to 4-12, Appendix A.

5. Mueller, Ivan I. and Ramsayer, Karl H. *Introduction to Surveying.* New York: Frederick Ungar Publishing Company, 1979, Chapter 5.

6. Stoughton, Herbert W. ''Introduction to Least Square Adjustment for the Land Surveyor'', *Proceedings of the American Congress on Surveying and Mapping,* October 18-21, 1977, pp. 375-427.

INDEX

Another book by Buckner . . .

A Manual on

ASTRONOMIC AND GRID AZIMUTH

R. B. Buckner, Ph.D.

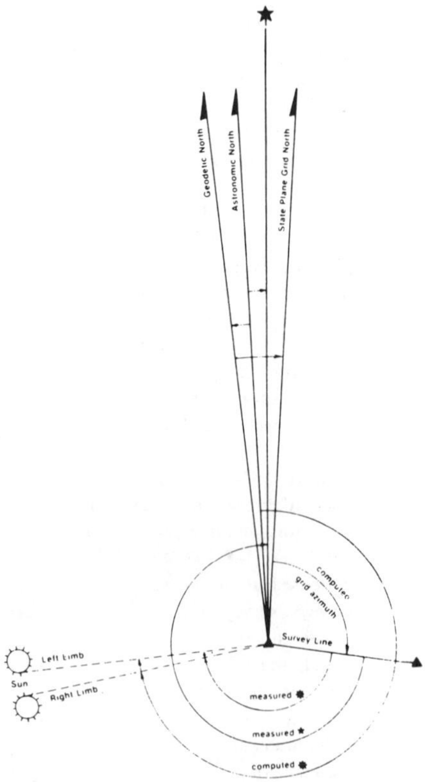

- Field Procedures
- Theory and Errors
- Laplace Corrections
- HP and TI Programs
- State Plane Zone Data

CONTENTS

This manual explains a fast and inexpensive method to determine astronomic azimuth to an accuracy of approximately ±5", using either Polaris or the sun and the same hour angle equation for either celestial body. Tables, maps, figures and other data and explanations provide for converting to state plane grid azimuth (in the U.S.) without the need to consult outside references. Errors and theory are thoroughly treated, but most of this is covered in appendixes, the main text focusing on the field and computational procedures. The manual is instructional in nature, with text explanations and examples presented in such a way that a full appreciation of the procedures and theory can be realized, the "button pushing" approach being avoided. A special emphasis is placed on the importance of, and the procedures to obtain the grid azimuth from astronomic. A chapter on state plane coordinates is also included. There is something in this manual for practitioners, students, professors, trainees studying for licensing, and anyone interested in the fascinating world of astronomy, measurement, and control systems. It is especially written for those needing or desiring to orient land surveys to a reliable reference meridian and to extend local survey control.